Fire Safety Aspects of Polymeric Materials

VOLUME 4
FIRE DYNAMICS AND SCENARIOS

Report of

The Committee on Fire Safety Aspects of Polymeric Materials

NATIONAL MATERIALS ADVISORY BOARD
Commission on Sociotechnical Systems
National Research Council

Publication NMAB 318–4
National Academy of Sciences
Washington, D.C.
1978

VOLUME 4 — FIRE DYNAMICS AND SCENARIOS

Fire Safety Aspects of Polymeric Materials

265 Post Road West, Westport, CT 06880

Printed in U.S.A.
Library of Congress Card No. 77-79218
I.S.B.N. 0-87762-225-6

NOTICE

The project that is the subject of this report was approved by the Governing Board of the National Research Council, whose members are drawn from the Councils of the National Academy of Sciences, the National Academy of Engineering, and the Institute of Medicine. The members of the Committee responsible for the report were chosen for their special competences and with regard for appropriate balance.

This report has been reviewed by a group other than the authors according to procedures approved by a Report Review Committee consisting of members of the National Academy of Sciences, the National Academy of Engineering, and the Institute of Medicine.

This study by the National Materials Advisory Board was conducted under Contract No. 4-35856 with the National Bureau of Standards.

The National Research Council was established in 1916 by the National Academy of Sciences to associate the broad community of science and technology with the Academy's purposes of furthering knowledge and advising the federal government. The Council operates in accordance with general policies determined by the Academy of authority of its Congressional charter of 1863, which establishes the Academy as a private, non-profit, self-governing membership corporation. Administered jointly by the National Academy of Sciences, the National Academy of Engineering, and the Institute of Medicine (all three of which operate under the charter of the National Academy of Sciences), the Council is their principal agency for the conduct of their services to the government, the public, and the scientific and engineering communities.

Printed in the United States of America.

FOREWORD

This volume is one of a series of reports on the fire safety aspects of polymeric materials. The work reported here represents the results of the first in-depth study of this important subject. The investigation was carried out by a committee of distinguished polymer and fire technology scholars appointed by the National Academy of Sciences and operating under the aegis of the National Advisory Board, a unit of the Commission on Sociotechnical Systems of the National Research Council.

Polymers are a large class of materials, most new members of which are man-made. While their versatility is demonstrated daily by their rapidly burgeoning use, there is still much that is not known or not widely understood about their properties. In particular, the burning characteristics of polymers are only now being fully appreciated and the present study is a landmark in the understanding of the fire safety of these ubiquitous materials.

In the first volumes of this series the committee has identified the limits of man's knowledge of the combustibility of the growing number of polymeric materials used commercially, the nature of the by-products of that combustion, and how fire behavior in these systems may be measured and predicted. The later volumes deal with the specific applications of polymeric materials, and in all cases the committee has put forth useful recommendations as to the direction for future actions to make the use of these materials safer for society.

Harvey Brooks, Chairman
Commission on Sociotechnical Systems

ABSTRACT

This volume in the series on the Fire Safety Aspects of Polymeric Materials focuses on improving fire safety by means of a better understanding of fire dynamics and extended use of fire scenarios. Specifically, this combined approach would yield: 1) the development of more fire-safe materials; 2) the design and development of more relevant materials test methods; and 3) the employment of more fire-safe designs.

The study reported in this volume assesses the state of the art of fire dynamics and recommends that the present decreasing trend of funding in this area be reversed and sharply increased if the full potential of the scientific approach to improved fire safety is to be realized.

Conclusions are drawn in each chapter and recommendations are made. The major conclusions and recommendations are extracted and combined in Chapter 2.

VOLUMES OF THIS SERIES

Volume 1	Materials: State of the Art
Volume 2	Test Methods, Specificiations, Standards, Glossary
Volume 3	Smoke and Toxicity
Volume 5	Executive Summary
Volume 6	Aircraft: Civil and Military
Volume 7	Buildings
Volume 8	Vehicles — Railed and Unrailed
Volume 9	Ships
Volume 10	Mines and Bunkers

PREFACE

The National Materials Advisory Board (NMAB) of the Commission on Sociotechnical Systems (CSS), National Research Council, National Academy of Sciences-National Academy of Engineering-Institute of Medicine, was asked by the Department of Defense, Office of Research and Engineering, and the National Aeronautics and Space Administration to "initiate a broad survey of fire suppressant polymeric materials for use in aeronautical and space vehicles to identify needs and opportunities, assess the state of the art in fire retardant polymers (including available materials, production, costs, data requirements, methods of test and toxicity problems), and describe a comprehensive program of research and development needed to update the technology and accelerate application where advantages will accrue in performance and economy."

In accordance with its usual practice, the NMAB convened representatives of the requesting agencies, and other agencies known to be working in the field, to determine how, in the national interest, the project might best be undertaken. It was quickly learned that wide duplication of interest exists. At the request of the other agencies with an interest in fire safety. Concurrently, the scope of the project was broadened to take account of the needs enunciated by the new sponsors as well as of those of the original sponsors.

The total list of sponsors of this study now comprises: Department of Agriculture, Department of Commerce (National Bureau of Standards), Department of Interior (Division of Mine Safety), Department of Housing and Urban Development, Department of Health, Education and Welfare (National Institute for Occupational Safety), Department of Transportation (Federal Aviation Administration, Coast Guard), Energy Research and Development Administration, Consumer Product Safety Commission, Environmental Protection Agency, Postal Service, as well as the original Department of Defense and National Aeronautics and Space Administration.

The Committee was originally constituted on November 30, 1972. The membership was expanded to its present status on July 26, 1973. The new scope was established after presentation of reports by liaison representatives covering needs, views or problem areas, current activities, future plans, and relevant resource materials. Tutorial presentations were made at meetings held in the Academy and during site visits, when the Committee or its panel met with experts and organizations concerned with fire safety aspects of polymeric materials. These site visits (upward of a dozen) were an important feature of the Committee's search for authentic information. Additional inputs on foreign fire technology were supplied by the U.S. Army Foreign Science and Technology Center and the Staff Scientist.

Dr. Herman F. Mark, Chairman

ACKNOWLEDGEMENTS

This report was originally drafted by the Fire Dynamics and Fire Scenario Panel comprised of the following members and liaison representatives of the Committee on Fire Safety Aspects of Polymeric Materials: Prof. Richard S. Magee (Panel Chairman), Dr. Raymond Friedman, Dr. Clayton Huggett, Rear Admiral William C. Hushing, USN (Ret.), Mr. Daniel F. Sheehan and Dr. George R. Thomas. It was then reviewed and finalized by the entire Committee. Conclusions and recommendations are the sole responsibility of the Committee. This volume was coordinated by Professor Magee.

A number of other people made substantial contributions to this volume. Some of them were colleagues of the Committee participants. They served unofficially, lending their ideas, advice, and assistance to accomplish various portions of the work. In particular, Prof. H. W. Emmons of Harvard University made a strong recommendation that fire dynamics be included in the Committee's scope and Dr. J. A. Parker of NASA-Ames Research Center, made a similar recommendation with regard to the use of scenarios. Official guests of the Committee who contributed tutorial presentations were Mr. S. martin, Stanford Research Institute; Profs. P. J. Pagni and R. F. Sawyer, University of California, Berkeley; Lt. A. orphanides, Army Foreign Science and Technology Center; Dr. L. J. Hillenbrand, Battelle Memorial Institute; Mr. C. Yuill, Southwest Research Institute; and Dr. J. de Ris, Factory Mutual Research Corporation.

Staff support rendered by the National Advisory Board was excellent. Miss Carolyn Tuchis provided continuous secretarial services, with her pleasant effective manner making all the difficult tasks seem easy. Dr. Robert S. Shane, NMAB Staff Scientist (later, Consultant) was an essential asset to the Committee. His assistance, guidance and encouragement were invaluable, and his untiring efforts greatly facilitated the successful completion of this work.

Prof. Israel Katz, Northeastern University, served as editor for this volume and his efforts greatly improved the presentation of material.

Dr. Raymond L. Smith
President
Michigan Technological University
Houghton, Michigan 49931

Dr. Morris A. Steinberg
Director, Technology Applications
Lockheed Aircraft Corporation
Burbank, California 91520

Dr. Giuliana C. Tesoro
Adjunct Professor
Massachusetts Institute of Technology
278 Clinton Avenue
Dobbs Ferry, New York 10522

Dr. John E. Tilton
Commodities Division
United Nations Conference on
Trade and Development
Palais des Nations
CH-1211 Geneva 10, Switzerland

Dr. John B. Wachtman, Jr.
Division Chief
Inorganic Materials Division
National Bureau of Standards
Washington, D. C. 20234

Dr. Max L. Williams
Dean, School of Engineering
University of Pittsburgh
Pittsburgh, Pennsylvania 15261

NMAB STAFF:

W. R. Prindle, Executive Director
J. V. Hearn, Executive Secretary

NATIONAL MATERIALS ADVISORY BOARD
COMMITTEE ON FIRE SAFETY ASPECTS OF POLYMERIC MATERIALS

Chairman

HERMAN F. MARK, Emeritus Dean, Polytechnic Institute of New York, Brooklyn

Members

ERIC BAER, Head, Division of Macromolecular Science, Case Western Reserve University, Cleveland, Ohio

ALFRED BROWN, Celanese Corporation, New York

JOHN M. BUTLER, Manager, Organic Polymer Synthesis, Monsanto Research Corp., Dayton, Ohio.

RAYMOND FRIEDMAN, Vice President and Scientific Director, Factory Mutual Research Corp., Norwood, Massachusett

ALFRED GILBERT, Manager, Organic and Polymer Chemsitry Branch, Chemistry Laboratory, Corporate R&D, General Electric Company, Schenectady, New York

CARLOS J. HILADO, Research Professor, Institute of Chemical Biology, University of San Francisco, California

RAYMOND R. HINDERSINN, Sector Manager, Research Center, Hooker Chemical Co., Niagara Falls, New York

WILLIAM C. HUSHING, Vice President, Kensington Management Consultants, Bath, Maine

FRANK E. KARASZ, Professor and Co-Director, Materials Research Laboratory, Polymer Science & Engineering Dept., University of Massachusetts, Amherst

TED A. LOOMIS, Professor & Chairman, Department of Pharmacology, School of Medicine, University of Washington, Seattle

RICHARD S. MAGEE, Professor, Department of Mechanical Engineering, Stevens Institute of Technology, Hoboken, New Jersey

ELI M. PEARCE, Head, Department of Chemistry, Prof. of Polymer Chemistry and Engineering, Polytechnic Institute of New York, Brooklyn

CHARLES F. REINHARDT, Director, Haskell Laboratory for Toxicology and Industrial Medicine, E. I. duPont de Nemours and Co., Newark Delaware

ARNOLD ROSENTHAL, Celanese Research Company, Summit, New Jersey

DONALD G. SMILLIE, Chief Structures Development Engineer, McDonnell Douglas Corp., Long Beach, California

GIULIANA C. TESORO, Adjunct Professor Massachusetts Institute of Technology, 278 Clinton Avenue, Dobbs Ferry, New York

NICHOLAS W. TSCHOEGL, Professor, Department of Chemical Engineering, Division of Chemistry and Chemical Engineering, California Institute of Technology, Pasadena

ROBERT B. WILLIAMSON, Associate Professor of Engineering Science, Dept. of Civil Engineering, University of California, Berkeley

Liaison Representatives

JEROME PERSH, Staff Specialist for Materials and Structures, ODDRE, Department of Defense, Washington, D.C.

ERNEST B. ALTKRUSE, Chief Department of Clinics, Moncrief Army Hospital, Fort Jackson, South Carolina

ALLAN J. MC QUADE, U.S. Army Natick Laboratories, Natick, Massachusetts

GEORGE R. THOMAS, Department of Army, Army Materials and Mechanics Research Center, Watertown, Massachusetts

HERBERT C. LAMB, Naval Facilities Engineering Command, Alexandria, Virginia

DANIEL PRATT, Naval Ships Engineering Center, Prince George's Center, Hyattsville, Maryland

GEORGE SORKIN, Naval Sea Systems Command, Washington, D. C.

JACK ROSS, Wright Patterson Air Force Base, Ohio

BERNARD ACHHAMMER, National Aeronautics and Space Administration, NASA Headquarters, Advanced Research and Technology Division, Washington, D.C.

JOHN A. PARKER, National Aeronautics and Space Administration, Ames Research Center, Moffett Field, California

ARNOLD WEINTRAUB, Energy Research and Development Administration, Washington, D.C.

DAVID FORSHEY, Bureau of Mines, Washington, D. C.

CLAYTON HUGGETT, Center for Fire Research, National Bureau of Standards, Washington, D. C.

LESLIE H. BREDEN, Fire Safety Engineering Division, Center for Fire Research, National Bureau of Standards, Washington, D. C.

IRVING LITANT, Department of Transportation (DoT), Systems Center, Kendall Square, Cambridge, Massachusetts

GEORGE BATES, JR., Aircraft Division, DoT Federal Aviation Administration, National Aviation Facilities Experimental Station, Atlantic City, N.J.

RALPH RUSSELL, Aircraft Division, DoT Federal Aviation Administration, National Aviation Facilities Experimental Station, Atlantic City, N.J.

PAUL W. SMITH, Aviation Toxicology Laboratory, Civil Aeromedical Institute, Oklahoma City, Oklahoma

ROBERT C. MCQUIRE, DoT Federal Aviation Administration, Washington, D.C.

THOMAS G. HOREFF, DoT Federal Aviation Administration, Washington, D. C. (Until September 1974)

DANIEL F. SHEEHAN, DoT, U.S. Coast Guard, Washington, D.C.

WILLIAM J. WERNER, Department of Housing and Urban Development, Washington, D.C.

DONALD L. MOORE, Department of Housing and Urban Development, Washington, D.C.

IRVING GRUNTFEST, Environmental Protection Agency, Washington, D.C.

NELSON GETCHELL, U.S. Dept. of Agriculture, Beltsville, Maryland

WILSON A. REEVES, Cotton Finishes Laboratory, U.S. Dept. of Agriculture, New Orleans, Louisiana

HERBERT W. EICHNER, Forest Service, U.S. Dept. of Agriculture, Forest Products Laboratory, Madison, Wisconsin

RICHARD E. WIBERG, National Institute for Occupational Safety and Health, Rockville, Maryland

JAMES RYAN, Consumer Product Safety Commission, Bethesda, Maryland

JURGEN KRUSE, Materials Research Division, Office of Postal Science and Technology, Rockville, Maryland

Technical Advisors

IRVING EINHORN, Flammability Research Center, Division of Materials Science and Engineering, Flammability Research Center, College of Engineering, University of Utah, Salt Lake City

LARRY ROSLINSKI, Ford Motor Company, The American Road, Dearborn, Michigan

LAWRENCE M. ENGLEMAN, Washington Metropolitan Area Transit Authority, Washington, D.C.

MAUREE W. AYTON, Science Information Analyst, Congressional Research Service, Library of Congress, Washington, D.C.

National Materials Advisory Board Staff

ROBERT S. SHANE, National Materials Advisory Board, National Academy of Sciences, Washington, D.C.

CONTENTS

CHAPTER 1

INTRODUCTION

1.1 Scope and Methodology of the Study

The charge to the NMAB Committee on Fire Safety Aspects of Polymeric Materials was set forth in presentations made by the various sponsoring agencies. Early in its deliberations, however, the committee concluded that its original charge required some modification and expansion if the crucial issues were to be fully examined and the needs of the sponsoring organizations filled. Accordingly, it was agreed that the committee would direct its attention to the behavior of polymeric materials in a fire situation with special emphasis on human-safety considerations. Excluded from consideration were firefighting, therapy after fire-caused injury, and mechanical aspects of design not related to fire safety.

The work of the committee includes: (1) a survey of the state of pertinent knowledge; (2) identification of gaps in that knowledge; (3) identification of work in progress; (4) evaluation of work as it relates to the identified gaps; (5) developoment of conclusions; (6) formulation of recommendations for action by appropriate, of the benefits that might accrue through implementation of the recommendations. Within this framework, functional areas were addressed as they relate to specific situations; end-uses were considered when fire was a design consideration and the end-uses are of concern to the sponsors of the study.

Attention was given to natural and synthetic polymeric materials primarily in terms of their nature, composition, structure, relation to processing, and geometry (i.e., film, foam, fiber, etc.), but special aspects relating to their incorporation into an end-use component or structure also were included. Test methods, specifications, definitions and standards that deal with the foregoing were considered. Regulations, however, were dealt with only in relation to end-uses.

The products of combustion, including smoke and toxic substances, were considered in terms of their effects on human safety; morbidity and mortality were treated only as a function of the materials found among the products of combustion. The question of potential exposure to fire-retardant polymers, including skin contact, in situations not including pyrolysis and combustion were addressed as deemed appropriate by the committee in relation to various end-uses.

In an effort to clarify the understanding of the phenomena accompanying fire, consideration was given to the mechanics of mass and energy transfer (fire dynamics). The opportunity to develop one or more scenarios to guide thinking was provided; however, as noted above, fire-fighting was not considered. To assist those who might use natural or synthetic polymers in components or structures, consideration also was given to design principles and criteria.

Accordingly, as the committee completes segments of its work, it plans to present its findings in the following five disciplinary and five end-use reports:

Volume 1	Materials — State of the Art
Volume 2	Test Methods, Specifications, and Standards
Volume 3	Special Problems of Smoke and Toxicity
Volume 4	Fire Dynamics and Scenarios
Volume 5	Executive Summary — Volumes 1 through 4
Volume 6	Aircraft (Civil and Military)
Volume 7	Residential, Non-Residential and Custodial Buildings
Volume 8	Railed and Unrailed Vehicles
Volume 9	Ships
Volume 10	Mines and Bunkers

Some of the polymer applications and characteristics are in the classified literature, and the members of the committee with security clearance believed that this information could best be handled by special meetings and addendum reports to be prepared after the basic report volumes were completed. Thus, the bulk of the output of the committee would be freely available to the public. Considering the breadth of the fire safety problem, it is believed that exclusion of classified information at this time will not materially affect the committee's conclusions.

1.2 Scope and Limitations of This Report

This report in the series on Fire Safety Aspects of Polymeric Materials specifically examines two different yet highly interrelated approaches to improving fire safety: fire dynamics and fire scenarios. Currently, neither of these approaches to the fire problem is sufficiently used to derive such potential benefits as the development of better materials and relevant test methods and the employment of improved fire-safe designs. For this reason this volume focuses on achieving improved fire safety by means of a better understanding of fire dynamics and the greater use of fire scenarios.

Fire dynamics is the scientific description of fire phenomena in quantitative terms. Phenomena of interest include: ignition, flame spread, fire growth, maximum burning intensity, products of combustion, movement of combustion products through a building or other structure, detection, extinguishment, and effects of fire on humans.

Fire scenarios are generalized descriptions of actual or hypothetical, but credible, fire incidents. Frequently, a scenario is based on one (or more) actual fire incidents; however, since in most real fires complete knowledge of all pertinent events is impossible to obtain, scenarios are generally based on a combination of fact and speculation. A scenario should include all details relevant to the development of a fire and the subsequent human behavior and protection device response.

Each scenario should be analyzed in depth to yield the maximum benefit. The general approach is analogous to the Case Study Method, which is used in other disciplines for the study of complex problems.

Where the goal is to improve fire safety, it may be necessary to select from a set of alternatives such as: replace existing materials with others having greater fire safety, but which are perhaps less suitable or more expensive; modify the existing design to reduce the hazard; increase the awareness of the users to hazards through education; control ignition sources, etc. To choose the most effective approach, an intimate knowledge of fire and fire behavior is required. An understanding of fire dynamics, coupled with the use of appropriate fire scenarios, can provide such knowledge.

An alternative to the detailed scenario approach is the statistical approach. For example, "Smoking, electrical faults, and cooking cause x, y, and z percentages respectively of all fatal fires in the home." In essence, such a statement could be considered to consist of three highly condensed scenarios. While such statistics are useful, they fail to provide the specific information required for decisions to replace, redesign, educate, or even tolerate a given hazard. This flaw in the statistical approach is best illustrated by examples as follows:

1. Introduction of a fire retardant into a polymeric material may reduce the probability of ignition, but doing so may increase the optical density of smoke produced if ignition does occur. Under what conditions does the use of a fire retardant lead to an increase in overall fire safety?

2. Material A is harder to ignite than material B, but, once ignited, spreads flame much more rapidly. Under what conditions is A more hazardous than B?

3. A given material will not ignite when exposed to a match flame for 5 seconds, but will ignite when exposed to a flaming waste-basket for 3 minutes. How safe is this material in a given application?

4. An item of modest size ignites and is consumed by flames, but there are no other flammable items nearby, so the fire goes out. Will a sufficient volume of toxic combustion products have been produced to cause fatalities?

To analyze each case, we clearly need to know the chemical nature of the combustibles, the geometry of the fire compartment and adjacent compartments, ventilation factors, the locations of occupants, the time lag until combustion products reach a human or automatic detector, as well as the possibilities of extinguishment, escape or rescue. Such cases require detailed scenarios.

Obviously, statistical information identifies the general types of fires which most commonly lead to fatalities (or major property loss), and the fire field certainly needs more and better raw statistics; but, to fully benefit from such information, we then need to select representative scenarios containing all the relevant details of a fire challenge and the related human response. It must also be recognized that statistics are not a good guide-post for certain classes of problems. For example, catastrophes of major proportions (high-rise buildings, wide-body aircraft) are infrequent and, therefore, statistically valid occurrences of events are not available; or a new product is introduced and years of experience are required for statistical loss data to accumulate to show its effect on fire safety. Even though actual fire scenarios may not be available in such cases, the generation of plausible hypothetical scenarios and their analyses are recommended.

The scenario approach helps clarify the nature and extent of hazards associated with products. In fact, a set of scenarios concerning a given product may permit better generalizations about the safety of that product than can otherwise be formulated. Moreover, scenarios are needed to establish the relevance of test methods. For example, if an expensive full-scale fire test is to be performed on a product, considerable guidance in planning the test could be obtained by prior scenario analysis.

The science of fire dynamics introduces a mixed set of benefits and problems relating to the development and interpretation of scenarios. For example, were the science of fire dynamics completely developed, all components of any fire scenario (except possibly those involving human response) could be predicted by a computer program. Unfortunately, while a significant body of fire-dynamics knowledge exists, it is highly incomplete. Many years of intensive research will be required to fill in the gaps, and the present level of research activity is inadequate. Yet, despite its incomplete status, the present knowledge of fire dynamics is sufficient to analysis of a scenario should include not only data as to what actually happened in a fire, but also speculation about alternative events which could easily have occurred under slightly differing circumstances. Existing fire dynamics knowledge is the only basis for postulating such variations of a scenario, short of performing realistic full-scale fire tests. Moreover, to reduce testing costs, it may be necessary to perform a "partial full-scale" fire test in which only one corner of a fire room is furnished realistically and the rest is bare. An analysis of the fire dynamics involved, utilizing current knowledge, should indicate the degree of validity for such a test. Obviously, since the relation of laboratory testing methods to realistic fire behavior involves fire dynamics, even our incomplete state of knowledge should be brought to bear to identify qualitatively, at least, such key parameters as radiant flux level, ventilation, etc. Preliminary evaluations of proposed design changes for fire-hardening also require current fire-dynamics inputs.

Fire dynamics and fire scenario generation and analysis are really two different specialties, each requiring personnel with somewhat different training and aptitude; combustion research engineers and scientists in one case and operations research specialists in other. Despite these differences in content and usage, the two areas of professional activity have been combined arbitrarily to illustrate their critical interrelationships.

In scope, this volume covers: the state of the art of fire dynamics and recommendations for its advancement; the need for the fire scenario approach to improve fire safety; generalized guidelines for developing and analyzing fire scenarios; some examples of fully developed and analyzed scenarios; and brief examples show how scenarios can be varied. Fire dynamics and scenario factors involving human response to fire are covered only briefly in this volume.

Statements in this volume must not be taken out of context and applied to other situations. This viewpoint must be emphasized lest information that appears in the published reports of this committee's study be misapplied.

CHAPTER 2

SUMMARY, CONCLUSIONS AND RECOMMENDATIONS

2.1 Summary

The physical and chemical aspects of fire render it a complex phenomenon. Over a period of several thousand years man has learned to use fire and, to some extent, control it, but many processes associated with ignition, growth, and extinguishment of fire remain largely unknown. Practical solutions for many fire situations have been devised without full knowledge of the fundamental processes involved and their interrelationships; however, many of these solutions have been based only on post-fire analyses and do not provide an acceptable level of fire safety in today's environment.

Modern technology provides us with larger and more elaborate living and transportation units that have interiors and furnishings made from new polymeric materials. Sufficient knowledge about the fire safety aspects of these materials is lacking and, as a result, the potential danger to life approach to fire prevention and control has become a necessity.

In each case, where a potentially hazardous polymeric material is identified, a choice must be made: replace with a more fire safe but usually more expensive or less satisfactory material; or minimize the hazard by design modification; or control potential ignition sources; or educate the users; or decide that the hazard is acceptable. To make such complex decisions wisely, intimate knowledge of fire behavior is required. Unfortunately such knowledge is incomplete. Our current knowledge needs to be greatly expanded. Better understanding of fire dynamics and increased use of appropriate fire scenarios can help significantly.

2.2 Conclusions and Recommendations

Designers and builders need a stronger basis for risk assessment and trade-off studies. An understanding of fire dynamics, supported by detailed and relevant scenarios, can provide this basis.

2.2.1 Fire Dynamics

Fire dynamics is the scientific description of fire phenomena (e.g. ignition, flame spread) in quantitative terms. It encompasses chemistry, physics, mathematics, fluid mechanics as well as heat and mass transfer. While fire dynamics is emerging as a scientific discipline, the complex nature of its fundamental processes, coupled with a relatively small funding effort[1], account for the rather slow progress in quantifying fire behavior. Despite the modest support of relevant scientific research, chiefly in the United States, England and Japan, a better understanding of important aspects of fire behavior is evolving. While a certain body of fire dynamics knowledge exists, it is in a highly incomplete state, and many years of intensive research are needed to fill in the gaps. Our current level of understanding of fire dynamics is not sufficiently developed to permit thorough scientific analysis of most fire test methods, fire scenarios or fire modeling procedures.

The behavior of a given material in a fire is dependent not only on the properties of the fuel, but also on the fire environment to which it is exposed. Consequently, if test methods are to be meaningful they should simulate the critical fire dynamic conditions. An understanding of fire dynamics is essential if such critical conditions are to be identified, and consequently, fire dynamics can be extremely useful in test method selection and development. In addition, valid modeling procedures, based on evolving fire dynamic principles, offer some promise for reduction in the costs of large scale fire tests as well as dependence thereon.

In line with the foregoing, the *Committee recommends* that, to obtain the full benefits of a scientific approach to improved fire safety, action be implemented as follows:

1. An enlarged program, of sufficient magnitude and stability to ensure an adequate level of sustained effort in the specific area of fire dynamics, should be established and funded by government.
2. This program should be conducted primarily by academic, non-profit, and governmental research organizations, but its individual projects should be closely monitored by advisory boards that include broad representation from manufacturers, users, and members of standards-setting organizations.
3. At least an additional 50 engineers and scientists (approximately equal to the current effort) should be supported over a 10-year period in the specific area of fire dynamics research. This addition of personnel would represent an increased cost of approximately $2.5 million per year, approximately 0.02% of the annual societal cost of fires in the U.S. (The President's Commission on Fire Prevention and Control estimate for 1972 of the cost of fires to society was some $11.4 billion).
4. In view of the shortage of qualified engineers and scientists in the appropriate disciplines, the program should be brought to full strength over a period of four years.
5. Fire dynamic principles should be included in the training of fire protection engineers, code officials, architects, design engineers, etc. Universities should be encouraged to include such material in appropriate graduate curricula. Additional educational efforts, e.g., short courses, should also be stimulated on a continuing basis.
6. Fire dynamics expertise should be employed when developing new test methods and for the validation and improvement of existing test methods.

2.2.2 Fire Scenarios

Fire scenarios[2] are generalized descriptions of actual or hypothetical fire incidents. Each scenario should include all details relevant to the development of a fire and subsequent behavior of people and/or automatic protection devices. Furthermore, each scenario should be analyzed in depth to yield maximum benefits. Although each fire is different, the development and analysis of fire scenarios leads to the identification of common elements in all fires, which in turn allows identification of the critical stages in fire development and suggests opportunities for prevention as well as control methods.

Scenarios should be prepared to permit generalization from each particular incident described. They should also provide a basis for exploring alternative paths of fire initiation and growth and for analysing the effects on fire safety performance of changes in material, design and operating procedures.

Scenario analysis consistently confirms that our fire safety development efforts are inadequately supported and, unfortunately, are largely directed toward items or components, when in fact, our problems are those of complex systems.

The application of fire scenario analysis appears to be the most productive method for identifying effective means to improve fire safety in our increasingly complex environment. Analyses of fire scenarios pin-point hazards, suggest opportunities to improve fire prevention, and direct attention to various methods for control. At present, insufficient use is being made of the fire scenario technique to advance fire safety.

In view of the large potential benefit to reduce fire hazards offered by analyses of appropriate fire scenarios, the *Committee recommends* that:

1. A wide spectrum of generalized fire scenarios, based on real or hypothetical, but credible, incidents be developed. The specific fire dynamic elements in these scenarios, e.g., rate of fire spread, rate of heat release, etc., should be further quantified as necessary by information obtained from large scale experiments. The ultimate aim of this recommendation is to develop the capability of employing analytical models to predict fire hazard and thus replace the need to conduct expensive large scale experiments. The ultimate aim of this recommendation is to develop the capability of employing analytical models to predict fire hazard and thus replace the need to conduct expensive large scale experiments.

2. Scenarios be used for the analysis of fire hazard and for the development of methods to provide increased safety. In particular, scenarios should provide bases for materials selection, formulation of design criteria, validation of test methods, promulgation of codes and standards, as well as the determination of research and development objectives.

3. The design and procurement of any system, subsystem or structure, employing polymeric materials, where fire safety is a design consideration, be based on the development and testing of the design against appropriate fire scenarios. This recommendation, in essence, requires the development of a fire impact statement.

4. Fire safety personnel be trained in the development and use of fire scenarios. A prime objective of such training is to enable them to identify critical fire hazard elements and determine appropriate protective measures.

---

[1]Total available funding in the U.S.A. for research on fire dynamics is estimated to have increased from $2.4 million in FY 1973 to $2.7 in 1976, and, as of February 1976, is scheduled to *decrease* to $2.0 million in FY 1977 (See Table 1).

[2]For a full discussion of scenarios, what they are, and how to use this important tool in various activities, please turn to 4.2 through 4.6. However, as with other analytical tools, one should not rely on a single source to organize one's thinking; all available information should be used by a decision maker.

CHAPTER 3

FIRE DYNAMICS

3.1 Introduction

Our knowledge of the complex physical phenomena comprising fire behavior — ignition, fire growth, smoke movement, extinguishment, etc. — was obtained primarily from mankind's long history of experience with fire, and is mainly qualitative rather than quantitative. This body of knowledge is far from optimum for dealing with the challenge of controlling fire risk associated with new materials, sometimes in new configurations. The development of a scientifically-based quantitative discipline of fire dynamics would be a powerful supplement to fire behavioral knowledge gained from post-fire analysis and full-scale testing.

this chapter reviews the current status of fire dynamics as a scientific discipline, indicates potential practical applications, nd recommends future efforts for the further development of fire dynamics.

If a material were expected to be exposed to fire risk in one and only one precisely defined set of circumstances (size, orientation, type of ignition source, method of applying ignition source, ventilation conditions, environmental conditions, etc.), it would be obvious how to test it for fire hazard. One must evaluate a series of candidate materials under a particular set of circumstances and note which materials are satisfactory or unsatisfactory. One would then evaluate smaller samples of the same series of materials by means of a proposed test method, which would become validated for future use if the results of the two procedures correlated.

Frequently, this idealized approach is not directly applicable because the material of interest may be exposed to fire risk in a wide variety of ways rather than in a simple well-specified set of circumstances. Thus, an incredibly large number of burn experiments would be required to explore fully all the permutations and combinations of the variables. A second reason is that in some cases even a few realistic burn experiments would not be feasible; for instance, if each experiment required destruction of a high-rise building or a Boeing 747 airplane, or, if an experiment required exposure of human beings to lethal fumes. As a result many fire test methods are considered to be highly unrealiable indicators of hazard. For many other cases, no appropriate test methods exist. Tests are particularly weak in regard to fire spread and growth, as well as noxious gas and smoke production.

If scientific studies of combustion were sufficiently advanced, one might hope to turn to scientists for the development of theoretical links between physical and chemical phenomena involved in test methods and the corresponding phenomena involved in real fires. Such theoretical knowledge, even if incomplete, would appear to be useful in generalizing from limited fire experience; it would be a powerful supplement to empiricism. However, combustion scientists have made rather limited progress in untangling the complexities of fire behavior and are unable at this time to analyze most fire situations in fundamental terms. Research, involving highly idealized materials and geometries, is progressing on a limited scale. In other research efforts, some realistic fires are highly instrumented to obtain useful information. However, much more effort is needed before we can expect major contributions by combustion scientists to more relevant test methods, the development of better materials, and more fire-safe designs.

3.2 The Scope of Fire Dynamics

What is fire dynamics? In broad terms, it involves quantitative descriptions of fire phenomena on a scientific basis. An understanding of these phenomena may be obtained on any one of four levels, as follows:

On the most basic level, we consider fire to be governed by a combination of effects tracing back to certain established fields of scientific study such as: the chemical thermodynamics and stoichiometry of combustion; the chemical kinetics of pyrolysis and combustion reactions; the transfer of combustion energy by conduction, convection, and radiation; as well as the motion of combustion gases as affected by buoyancy, thermal expansion or mechanical force, and as modified by constraining walls, viscous effects, inertia, turbulence, the properties of hot gases, etc. These factors are of interest for application purposes in addition to fire dynamics, but are considered to be proper subjects for academic rather than industrial research. Knowledge bearing upon these factors constitutes the foundation of fire dynamic studies.

On the second level, we may break fire down into a series of phases or stages, and consider fire dynamics to be an analysis of any of these phases:

- smoldering
- spontaneous ignition
- piloted ignition
- horizontal or downward flame spread over solids (thermally thin or thermally thick)
- upward flame spread over solids
- flame spread over a liquid below its flash point
- flame spread over a liquid above its flash point
- burning rate of a liquid pool
- burning rate of a solid slab
  - (charring or non-charring)
  - (vertical or horizontal, facing up or down)
  - (laminar or turbulent flame)
- formation of toxic species in a diffusion flame
- formation of aerosols in a diffusion flame
- radiation emitted by a diffusion flame
- extinguishment by heat loss
  - (radiation, convection, heat-absorbing substances)
- extinguishment by reduction of oxygen
- extinguishment by chemical inhibitors

Research in these phases is conducted largely but not entirely in academic institutions. Results would be of general interest to persons beyond the fire community concerned with combustion such as furnace designers and air pollution control administrators.

On the third level, we consider fire dynamics to be concerned with complex processes generally involving interactions of two or more of the phases given in the foregoing list. Some examples of such fire dynamics are as follows:

- burning rate of an object as influenced by radiative feedback from the environment
- burning rate of an object in a non-combustible compartment as influenced by ventilation of the compartment
- generation of incomplete combustion products as influenced by either of the above burning rate conditions
- mutual interactions of two adjacent burning objects
- spontaneous ignition of pyrolysis gases from a hot object as influenced by turbulent free convection and mixing
- properties of smoke from a fire in a compartment as influenced by the mixing, cooling, and "aging" or agglomeration, which occurs in the interval between smoke generation and its arrival at a detector station
- adsorption of toxic gases from fires by aerosols from the same fires, as cooling occurs
- analysis of radiant emissions, transmission, and absorptions in a compartment at a pre-flashover stage of fire
- the effect of physical scale on fire turbulence, on fire radiation, and ultimately on fire behavior
- delineation of relative effects of chemical kinetics and physical factors on a fire near an extinguishment conditions
- determination of the source of undesirable products of incomplete combustion in a fire; i.e., surviving initial pyrolysis products vs. products formed in gas-phase reactions in or near a flame
- effects of a long-term exposure of materials to various ambient conditions on subsequent fire behavior
- identification of the mechanisms involved in the interaction of water spray with a fire, e.g., cooling of pyrolyzing solids cooling and/or diluting of flame with steam, prewetting of adjacent fuel, absorbing radiation, entering into the flame chemistry via $C + H_2O$ or $CO + H_2$, entraining air with the water spray, as well as exploding superheated droplets in molten polymer and splattering fuel that feeds the combustion process.

Obviously, successful fire dynamics studies along these lines would strengthen the engineering judgement needed to validate test methods for the realistic determination of fire hazards. Academic researchers tend to avoid this third level of problems, many of them feel there are still so many major unknowns in the first and second level problems that third level of problems are too difficult to treat properly.

So far, except for mentioning toxicity, we have been discussing fire as if it occurs in an uninhabited world. Consequently, the fourth level of fire dynamics involves interactions of fire phenomena with human response. Problems such as the following exist:

- sensory detectability of fire (including smell and sound)
- vision as affected by smoke and/or lachrymatory gases
- panic or confused thinking as induced by fire phenomena
- burn damage of skin, particularly as influenced by the rate of flame spread, melting-dripping, etc.
- toxicity, including combined (synergistic) or sequential effects of various toxic compounds
- toxicity, including prefire condition of victim (blood-alcohol content, heart or circulatory disease, etc.)
- human ability to control fire at various stages of development as governed by training, equipment available, panic, etc.

It is recognized that the study of human behavior may well be considered as outside the scope of a treatise on polymeric material behavior, but it seems reasonable to call attention to this important aspect of the problem to complete the catalog of what we need to know to obtain a complete understanding of fire dynamics.

While comprehensive, the foregoing four levels of approach to fire dynamics exclude phenomena related to the post-flashover development of a fire. They do not consider the spread of a fire from one part of a building to another; the responses of building components (ventilation system, windows, elevators, etc.) to a fire; the spread of fire from one structure to an adjacent structure; and technology concerned with detection, communication, escape procedures, etc. Although a strictly scientific study of some elements of these factors is possible; engineering or systems analysis, involving largely empirical and state-of-the-art knowledge, is primarily required. Accordingly, the committee feels that study of such complex processes should not be called fire dynamics, even though such studies should utilize fire dynamics when possible.

3.3 Current State of the Art of Critical Fire Dynamic Elements

3.3.1 Introduction

In recent years, most of the current research on fire dynamics in the U.S. has been funded to various grantees by the Research Applied to National Needs (RANN) program of the National Science Foundation (NSF). However, the Federal Fire Prevention and Control Act (FFPCA) of 1974 assigned responsibility for "basic and applied research for the purpose of arriving at an understanding of the fundamental process underlying all aspects of fire" to Center for Fire Research (CFR) of the National Bureau of Standards (NBS). The Center conducts an internal program and will provide much of the future support for fire dynamics research[1]. Several other government agencies support some fire research which includes, to a minor degree, fire dynamics studies. Factory Mutual Research Corporation (FMRC) supports its own internal program of fire dynamics research. Table 1, based on estimates, is a representation of the total funding for fire dynamics research. (In Table 1, research on the toxicology of fire gases is not included.)

Table 1. Fire Dynamics (FD) as Budget in U.S. Programs.

(Millions of Dollars)

FISCAL YEAR	NSF-RANN FIRE PROGRAM	NFPCA RESEARCH PROGRAM	NSF + NFPCA TOTAL	%FD*	FD	NBS CFR TOTAL	%FD*	FD	FMRC-INTERNAL PROGRAM TOTAL	%FD	FD	MISC. FD*	TOTAL FIRE DYN. RESEARCH
73	2.00	0	2.00	60%	1.20	1.9	10%	0.19	1.0	50%	0.50	0.50	2.39
74	1.70	0	1.70	60%	1.02	3.1	10%	0.31	1.0	50%	0.50	0.50	2.33
75	0.65	1.12	1.77	60%	1.06	3.1	10%	0.31	1.1	50%	0.55	0.50	2.42
76	1.16	1.20	2.36	60%	1.42	3.1	10%	0.31	1.0	50%	0.50	0.50	2.73
77	0	1.50	1.50	20%	0.30	4.5	16%	0.72	1.0	50%	0.50	0.50	2.02

*Estimated.

Table 1 shows that the total available fire dynamics funding increased from $2.39 million in FY 1973 to $2.73 million in FY 1976, but, as of February 1976, is scheduled to decrease to $2.02 million in FY 1977.

Table 2 shows the major U.S. research organizations performing fire dynamics research. In about half the cases, however, the activity consists of a single professor working part-time with one or two graduate students.

A rather complete (although not quite current) guide to the published literature relevant to fire dynamics is *Fire Research Abstracts and Reviews,* published several times each year since 1958 by the National Academy of Sciences. Another source is *References to Scientific Literature on Fire,* published annually for many years by the Joint Fire Research Organization, Borehamwood, Herts., England. The following journals contain occasional relevant papers: Combustion and Flame, Combustion Science and Technology, Journal of Fire and Flammability, and Fire Technology. See also the biennial international Symposia on Combustion (The Combustion Institute, Pittsburgh). A new journal, Fire and Materials, commenced publication in 1976. Finally, some relevant books are referenced (3.1-1, 3.1-2, 3.1-3, 3.1-4).

Table 2. Some U.S. Research Organizations Performing Fire Dynamics and Related Research.

(in roughly approximated decreasing order of level of activity)

1. Center for Fire Research, National Bureau of Standards, Gaithersburg, Md., Dr. J. Lyons, Dr. R. S. Levine, Dr. J. Rockett, Dr. C. Huggett, Dr. J. Quintiere, Dr. T. Kashiwagi, Dr. H. Baum, Dr. R. J. McCarter.
2. Factory Mutual Research Corporation, Norwood, Mass., Dr. R. Friedman, Dr. J. de Ris, Dr. G. Heskestad, Dr. G. H. Markstein, Dr. R. L. Alpert, Dr. P. A. Croce, Dr. A. Modak, Dr. F. Tamanini, Dr. A. Tewarson.
3. University of California, Berkeley, Cal., Professor R. B. Williamson, Professor C. L. Tien, Professor P. J. Pagni, Professor R. Sawyer.
4. University of Utah, Salt Lake City, Utah, Professor J. D. Seader, Professor N. W. Ryan, Professor I. Einhorn.
5. Applied Physics Laboratory, Johns Hopkins University, Silver Spring, Md., Dr. W. G. Berl, Dr. R. M. Fristrom.
6. Harvard University, Cambridge, Mass., Professor H. W. Emmons.
7. Massachusetts Institute of Technology, Cambridge, Mass., Professor T. Y. Toong, Professor J. A. Fay, Professor G. C. Williams.
8. Princeton University, Princeton, N. J., Professor I. Glassman, Professor M. Summerfield, Professor W. A. Sirignano, Dr. F. L. Dryer.
9. Georgia Institute of Technology, Atlanta, Ga., Professor B. T. Zinn, Professor P. Durbetaki.
10. University of Notre Dame, South Bend, Ind., Professor K. Yang, Professor J. R. Lloyd, Professor M. L. Doria.
11. University of California, San Diego, Cal., Professor F. A. Williams.
12. Pittsburgh Mining and Safety Research Center, U.S. Bureau of Mines, Pittsburgh, Pa., Dr. D. Burgess.
13. Brown University, Providence, R. I., Professor M. Sibulkin.
14. Cornell University, Ithaca, N. Y., Professor K. Torrance.
15. Stevens Institute of Technology, Hoboken, N. J., Professor R. S. Magee.
16. University of Montana, Missoula, Montana, Professor F. Shafizadeh.
17. Northwestern University, Evanston, Ill., Professor M. C. Yuen.
18. California Institute of Technology, Pasadena, Cal., Professor E. Zukos.
19. State University of New York, Stony Brook, N. Y., Professor R. Lee, Professor A. L. Berlad.
20. University of Maine, Orono, Maine, Professor A. Campbell.
21. University of Washington, Seattle, Wash., Professor R. Corlett.
22. Pennsylvania State University, State College, Pa., Professor G. Faeth.

To illustrate the current state of the art, ignition and flame spread will be reviewed; then several examples of other fire-dynamic studies will be discussed briefly.

3.3.2 Ignition

The ignition of a combustible material is the first step in any fire scenario and, therefore, is important to fire prevention. Furthermore, once a fire starts, the ignition delay times of other materials, coupled with flame spread, will affect the rate at which the fire spreads and develops. hence secondary ignition of other materials is also important to fire development.

Most polymeric materials ignite and burn if they are sufficiently heated in the presence of air or other atmosphere containing sufficient oxygen. Consequently, from the viewpoint of fire safety, it is desirable to know how long it takes a particular polymer to ignite under various fire conditions, e.g., oxygen concentration, environmental gas temperature, heating rate, etc. This section reviews what is known about polymer ignition.

Historically, modern ignition theories have evolved from research directed towards determining the ignition mechanisms of coal and solid propellents (3.2-1)[2]. Early attempts at formulating an ignition theory by Frank-Kamenetskii resulted in a bulk thermal explosion theory (3.2-2). The first sophisticated attempt at a "surface-ignition" model was made by Hicks (3.2-3). His analysis solved the solid-phase energy equation, including a term that allowed for exothermic chemical heating in the solid which had an exponential dependency of rate on temperature, for the transient surface heating case. Gas-phase processes were not considered. Since, for most polymeric materials under rapid heating conditions, chemical heating in the solid has not been established as being important during ignition, Hicks' analysis may not be generally applicable. Subsequent attempts to employ a purely solid-phase thermal theory to describe heating of a olid fuel to an "ignition temperature", have resulted in many solutions to modified forms of the solid-phase energy equation (accounting for rate of energy convection due to regression of the surface, and rate of absorption of energy in depth due to optical transparency of the solid) for a wide variety of boundary conditions and heating conditions (3.2-4, 5, 6, 7, 8). some of these solid-phase thermal theories have been successful in correlating ignition data for cellulosic materials in particular (3.2-9).

Two conspicuous deficiencies of solid-phase thermal ignition theories are: 1)their insensitivity to conditions in the surrounding gas-phase environment except as they affect thermal properties of the gas and the ensuing solid-phase surface heat loss, and 2) the absence of any recognized condensed-phase exothermic or endothermic reactions in many fuels. These deficiencies led to the development of a gas-phase ignition model in the late 1950's by workers at Princeton (3.2-10). This gas-phase ignition model received greater attention during the next 15 years, resulting in numerous improvements and increased sophistication of the model (3.2-11, 12, 13, 14). Observations by Simms, discussed later, of the radiant ignition of cellulose, confirm ignition initiating in the gas phase. The physical processes occurring in the gas-phase model are described below.

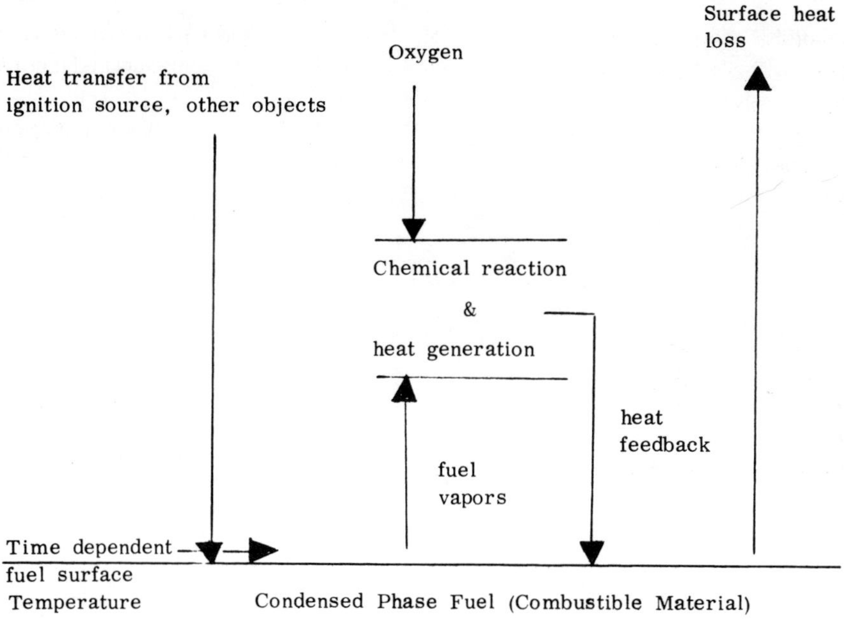

Figure 1. Physical process in gas-phase ignition model.

In the physical model, illustrated in Figure 1, the process of ignition begins when the solid fuel surface starts to heat up as a result of heat transfer from the adjacent surroundings. (This heat transfer may be by conduction, convection, and/or radiation). As the solid fuel surface temperature increases in response to this heat input, fuel vapors are emitted from the surface. The fuel vapors mix by convection and diffusion with oxygen in the adjacent boundary layer, ultimately leading to heat liberation by chemical reaction in the gas if the gas temperature is high enough. Some of this liberated heat may raise the surface temperature further and accelerate the process.

Thus, the gas phase ignition model has three characteristic times, the sum of which equals the time for ignition, or the ignition delay. These three periods are: 1) the thermal induction period, 2) the diffusion induction period, and 3) the chemical induction period. The thermal induction period is the time during which the solid fuel temperature is being raised by external heating to a temperature at which pyrolysis of the fuel begins. During the diffusion induction period, fuel gases are being evolved at the surface and diffuse outward into the oxidant. Finally, during the chemical induction period, the fuel and oxygen react exothermally. Therefore, delaying the onset of ignition implies lengthening one or more of these periods.

It is interesting to note at this point that in those ignition situations where the thermal induction period is much longer than the diffusion and chemical induction periods (e.g., low heating rates, high oxygen concentration), the thermal induction period "controls". Thus, it is not surprising that some ignition data have been correlated reasonably well by solid phase thermal theories, which model only the fuel bed thermal response and assume a fixed "ignition temperature".

Thus, most ignition situations involve heat and mass transfer in addition to polymer and combustion chemistry. In the gas-phase ignition model, the solid phase fuel pyrolysis rate (m) is assumed to be governed by a first-order Arrhenius expression:

$$m = \rho_s A_s e^{-E_p/RT},$$

where ρ_s = solid fuel density, A_s = pre-exponential factor of fuel pyrolysis rate and E_p = activation energy of fuel pyrolysis. The gas phase chemical reaction between fuel factors and oxygen is often assumed to be first-order with respect to both fuel and oxidizer and the rate of the reaction dependent upon the local temperature and reactant concentration as follows:

$$\text{rate} = C_F\, C_o\, A\, e^{-E_R/RT},$$

where C_F and C_o are the local concentrations of fuel and oxygen respectively and E_R = activation energy of the gas-phase reaction.

The resulting partial differential equations and boundary conditions describing the gas-phase ignition model, even in the one-dimensional case, are coupled and nonlinear. Analytical solutions are impossible, even though an asymptotic analysis (3.2-12) shows that ignition delay is dependent on parameters such as the mole fraction of oxygen. Therefore, the only approach to solving these quations is by computer analysis (3.2-11, 14). These solutions indicate the dependency of ignition delay on such factors as: oxygen concentration in the environment, ambient and local pressures, fuel pyrolysis rate and inert diluent in the environment for various assumed combinations of heating rates, preexponential factors, activation energies, etc. The results compare favorably with the little experimental data available and indicate that it is reasonable to expect ignition delays of the order of 2—10 seconds, even with relatively hard-to-degrade polymers, in air, at one atmosphere, when the gas temperature is 1200 °K (1700 °F) or more.

In some experiments and tests, one is concerned with *piloted* ignition rather than *spontaneous* ignition. The model described above is essentially for spontaneous ignition since it employs the feature of a runaway chemical reaction in the gas phase leading to ignition. In piloted ignition, the flame may not sustain itself, except perhaps momentarily, once the source of ignition energy is withdrawn or exhausted. Hence, it is instructive to examine the principal factors, involved in the ability of a just-ignited solid to continue burning after the ignition energy flux is removed, as follows:

I. The pyrolysis gases must issue from the surface into a boundary layer at a high enough rate to permit establishment of a flame at a position which is far enough from the surface to avoid quenching.

II. The flame must heat the surface at a rate sufficient to continue production of pyrolysis gases.

III. The rate of heat conduction (and possibly radiation) from the surface into the cold interior of the solid must not be excessive. This rate decreases with time, as the solid heats up.

IV. The rate of radiant heat loss from the surface outward to the colder surroundings must not be excessive.

V. The radiant input to the surface from more distant flames, or from hot gases and smoke, or from other facing hot surfaces, can compensate for loss mechanisms.

In special cases, additional factors may be present which affect the energy balance at the surface, including: melting and dripping, heat loss to adjacent cold inert surfaces, cooling by impinging water droplets from a sprinkler, exothermic oxidative reactions in the surface of the polymer, heat involved in $M_n \longrightarrow nM$ reaction (depolymerization). Therefore, even in the case of piloted ignition, heat and mass transport are important processes.

The scope and accuracy of experimental ignition delay data are not sufficient to test adequately the various theories, or to determine the relative importance of various processes to the ignition delay. Theories, by themselves, often do not provide quantitative predictions relevant to experiments. Moreover, the bases for the formulations of theories are sufficiently different that the theories cannot all be tested by the same experiment. Thus, while the gas-phase ignition model seems to best represent the ignition of polymeric materials, total experimental verification seems impossible. However, much useful empirical information about polymer ignition is available and some of the more important findings are as follows:

Flaming will not *persist* in a single wood slab not subjected to supplemental external heating unless the *average* temperature within the slab is greater than bout 320 ° C. The corresponding pyrolysis gas flux from the surface necessary for persistent burning must be greater than about 2.5 X 10^{-4} g/cm^2/sec. *Transient* flaming can occur when the *surface* temperature reaches 450 °C (convective heating), or 300 to 410 °C (Radiant heating in a quiescent ambient atmosphere).

Under ideal conditions, the *time* required to heat a wood slab is given thickness to a condition of sustained burning is proportional to $C^{\rho}x$ if the slab is "thermally thin," and to x^2/ω if the slab is "thermally thick," where x is slab thickness (or half-thickness if appropriate), C is specific heat, $^{\rho}$ is density, and α is thermal diffusivity. "Thermal thickness" is satisifed when the slab is substantially thicker than $\lambda(T^*-T_0)/I$, where λ is the thermal conductivity of the slab, T* is the "critical surface temperature for sustained ignition," and I is the incident heat flux on the surface. Thus, the dividing line between thermally thin and thermally thick behavior depends on the heating rate I.

In a recent study, Rangaprasad et al. (3.2-15) correlated radiant piloted ignition data for cotton and other thin cellulosics, and showed that all data could be correlated in a plot of t_i/W vs al, where t_i is time to ignition, W is fabric weight per unit area, and a is the absorptivity for incident radiation of intensity I. The minimum value of al is about 1.25 W/cm^2.

This discussion represents the highlights of what has been established concerning piloted ignition of cellulosic solids. The 320 °C average temperature criterion, as well as the 1.25 w/cm_2 absorbed radiant flux criterion, while useful, are purely emprical, and no attempt has been made to derive these numbers from the five fundamental factors listed above (pyrolysis gas concentration in boundary layers, heat transfer from flame to surface, internal heat transfer, radiant loss from surface, radiant flux from surroundings to surface). Such a development is held up not only because of mathematical complexity, but also, and primarily, because of the lack of quantitative knowledge of component processes involved. Some areas requiring further knowledge are:

1. thermal conductivity of partially decomposed wood
2. kinetics and energetics of primary pyrolysis
3. secondary reactions of pyrolysis gases and possibly oxygen in the char layer
4. flammability limit of pyrolysis gas mixtures in boundary layers
5. convective and radiative components of heat transfer from flame to surface
6. radiant emissivity of surface, which may change as char develops
7. initial oxidative reactions of pyrolysis gases in the gas phase

Upon turning its attention to various thermoplastics other than wood or cellulose products, the committee finds that very little systematic research on piloted ignition has been done. Thin thermoplastics, such as fabrics, are either ignited or locally destroyed by melting within a second or so, when a small flame is applied. The "limiting oxygen index" test shows that many thermoplastics in candle form can be ignited and will sustain combustion in atmospheres containing appreciably less than 21% oxygen, while sustained candle-like burning of a small wood stick requires as much as 28% oxygen.[3] It may be speculated that this difference is primarily due to relative magnitudes of radiative loss from the burning surface; the plastic surfaces being at roughly 400 °C, whereas the charred wood is at approximately 600 °C. Since loss varied as T_s^4, then the 200 °C difference in T_s at these temperature levels would account for three times as much radiant loss rate.

While it may be difficult to maintain sustained burning of a single semi-infinite slab of wood facing a cold ambient, it is easy to do so for a semi-infinite slab of many thermoplastics. Given a thick slab of polymethyl methacrylate, with edges and corners masked, the requirement for piloted ignition of the surface to lead to sustained burning is that a thin surface layer (a few mm thick) must be preheated to about 270 °C. Once such preheating has been accomplished, application of small pilot flame for a second or so will produce sustained burning.

Extensive experimental data has been reported for surface heating leading to spontaneous ignition of *cellulosic* materials; cf. Simms (3.2-16), Martin (3.2-17), Akita (3.2-18). The primary finding is that transient flaming ignition will occur once the surface reaches 500 to 650 °C. Ignition is observed to start in the gas phase rather than at the surface. The larger the heated surface area, the more readily ignition occurs. Simms finds that the occurrence of turbulence in the stream of gases emitted from the surface is important to the onset of ignition. He obtained photographs showing ignition to occur at the laminar-turbulent transition point in the flow field. Radiant surface heating, in the presence of cold ambient air, requires heating to a higher surface temperature than with convective heating by hot air, as would be expected.

There are fewer studies of spontaneous ignition of *plastics* under precisely controlled conditions. Kashiwagi et al. (3.2-19) ignited a copolymer of butadiene and acrylic acid in a shock tube containing O_2-N_2 mixtures within a few milliseconds, by using a fast-flowing oxidant atmosphere shock-heated to about 2000 °K. High-speed photographs showed ignition to occur in the gas rather than at the surface. The lower the O_2 content the further downstream the ignition occurred. These authors were not able to rule out the possibility that surface oxidation as well as gas-phase reaction were contributing to ignition, although they favored the purely gas-phase mechanism. Ohlemiller and Summerfield (3.3-13) ignited polystyrene and epoxy samples with a laser. Using fluxes of 90 to 400 watts/cm^2, they obtained ignition in 20 to 200 milliseconds in various O_2/N_2 atmospheres. The addition of carbon black to polystyrene very markedly reduced the ignition delay, presumably by causing the energy to be absorbed near the surface rather than in depth. The results could be described by an energy balance equation requiring the surface to be heated to about 650 °C as long as the O_2/N_2 ratio is high enough to prevent a diffusion limitation to the ignition reaction.

3.3.3 Flame Spread

The propagation of a flame over a combustible solid is an extremely complex process. However, since the flame spread rate is readily measured, a large body of experimental data involving flame spread rate has been obtained over the past decade. In spite of these data, there still exists uncertainty as to which physical and chemical parameters exert dominant effects.

Since heat must travel ahead from the flame to unignited material in order to propagate the flame, certain heat transfer modes must be involved. Yet, the relative importance of conduction or convection in the gas phase, conduction in the condensed phase, and radiation in the gas phase with regard to flame spreading rates, is not known even for the simplest cases. The motion of gases at the leading edge of the flame is of potential importance. Gas-phase chemistry including oxidant concentration, inert diluent, and pressure, as well as inter-diffusion of reactants, must also be considered. Moreover, the influence of solid phase surface chemistry, thermal properties and geometry cannot be ignored.

Results from many experimental investigations have been summarized by Friedman (3.3-1) and by Magee and McAlevy (3.3-2). They indicate that the flame spread rate was affected by many parameters as follows:

Physical and geometrical parameters:

- orientation of surface
- direction of propagation
- thickness of specimen
- surface roughness
- presence of sharp edges

initial fuel temperature
environmental pressure
velocity of environment
external radiant flux
humidity
specimen size

Chemical parameters:

composition of solid
composition of atmosphere

Most of the studies reviewed (3.3-1, 3.3-2) were conducted with horizontal or downward flame propagation, even though the very important upward spread process occurs several orders of magnitude faster and appears to be controlled by different processes. Since these reviews, further studies seeking to pinpoint the details of the horizontal or downward spread mechanism have been carried out by Hirano et al (3.3-3, 4, 5), Sibulkin et al. (3.3-6, 7, 8), Moussa et al. (3.3.9) and Fernandez-Pello and Williams (3.3-10). The principal controversies associated with these studies concern the contributions of the various modes of heat transfer at the leading edge of a flame, which may be different for thick vs. thin materials, cellulosic vs. thermoplastic materials, etc. Also for the first time, the details of the gas motion at the leading edge of the flame have been explored.

In order to provide a better understanding of the relative importance of the various parameters on flame spread phenomena, the effects of some of the physical and chemical parameters listed above are discussed in detail below.

Angle of inclination

The results for both polymethyl methacrylate and cellulose acetate indicate that there is little difference between horizontal flame spread rate and vertical downward spread rate over thick specimens. Hirano et al. (3.3-3, 4) explored the flame spread mechanism over thin paper sheets and found downward spread rates in the rate of 90° to 30° from the horizontal to be essentially constant; however, in the range of 30° to 0° to the horizontal, the spread rate seemed unstable and increased by repetitive acceleration and deceleration. At these smaller angles, the spread process seems to be controlled by the flame on the underside of the thin specemin. Thus, one must be careful not to generalize; the flame spread mechanisms can change when such thin specimens are employed and their inclinations are varied.

The much-more-rapid upward spread process has been studied very little. The flame spread rate increases with the angle of orientation (3.3-2, 3); increasing by an order of magnitude from the horizontal to vertical orientation. In fact, the rates become somewhat erratic and it is difficult to obtain reproducible results. Hansen and Sibulkin (3.3-11) made a preliminary study of spread up a small thermoplastic wall. Orloff et al. (3.3-12, 13) made a much more detailed study of 2-dimensional turbulent flame spread up thick vertical PMMA walls 157 cm and 356 cm high. Results of these studies show that the spread process is continuously accelerating, even at these heights, and that the spread rate is controlled by the largely radiative heat transfer from the upper part of the turbulent flame to the portion of the wall above the pyrolysis front.

Markstein and de Ris (3.3-14) have studied 2-dimensional upward spread with cotton fabrics held vertically and at various inclinations. The turbulent upward spread over fabrics should differ from the previously described behavior; as the flame climbs the fabric burns out, thereby limiting the ultimate size of the flame and hence the spread rate. With fabrics up to 150 cm long, the flame spread was found to continually accelerate, acquiring a turbulent character after a brief laminar period. Analysis showed that an asymptotic rate would have eventually been reached if the fabric sample had been longer.

Thus, the specimen inclination and direction of propagation exert a powerful influence on the flame spread rate. Further discussion of this subject is provided by Backer et al. (3.3-26).

Thickness of Specimen

Flame spread rates have been found to be independent of fuel bed depth for depths greater than approximately 0.25 cm, but they are inversely proportional to thickness for thinner sheets. This effect is rationalized using the concept of effective thermal thickness of the specimens. A "thermally thick" fuel bed is one in which thermal diffusion into the solid beneath the leading edge of a spreading flame is much less than the fuel-bed depth. A "thermally thin" fuel-bed is one in which subsurface temperature gradients are negligible. Whether or not a specimen is thermally thick or thermally thin depends on the fuel bed density, its thermal conductivity and specific heat, as well as the heating rate to the solid from the flame. This heating rate is a function of the spread rate and the preheat distance ahead of the flame.

The effective thermal thickness of a fuel bed strongly influences flame spreading phenomena. However, the effects of environmental pressure, oxygen mole fraction, and initial fuel bed temperature on the spread rate differ for thermally thick and thermally thin specimens.

Surface Roughness and Exposed Edges

The physical nature of a polymer can strongly affect the flame spread rate. A flame propagates approximately five times as fast over a smooth horizontal surface where edges are exposed as when they are inhibited (3.3-2). Moreover, exposed edges increase data scatter. Generally, the rougher the fuel bed surface the faster is the flame propagation rate.

Environmental Pressure

Flame spread rate has been shown to vary with pressure in accordance with the relationship:

$$V \sim P^N ,$$

where V = flame spread rate, and P = environmental pressure for a variety of materials over large ranges in pressure as well as at various orientations (3.2-2, 15). For flame spread over the top surface of horizontal thermally thick specimens, and for vertically downward spread over thermally thick specimens, the exponent $N \cong 0.75$. However, for thermally thin specimens and vertical downward spread, $N \cong 0.1$. For other orientations, or for specimens in the transition thickness between thermally thin and thermally thick, N falls somewhere between these two values.

Initial Fuel Temperature

Increasing the initial fuel bed temperature increases the flame spread rate, but there seems to be only limited data available on this effect. Early work (3.3-2), indicated a difference between the behavior of PMMA and cellulosic specimens attributed to an increase in initial temperature. Theoretical analyses yielded the prediction that

$$V \sim (T_b - T_o)^{-m},$$

where V = spreading rate, T_b fuel "burning temperature" (maximum surface temperature at leading edge of the flame), T_o = initial temperature, m = 2 for thermally thick fuel beds, and m = 1 for thermally thin fuel beds. Early work with PMMA confirmed the m = 2 exponent for thick specimens; however, the results with cellulosic specimens did not support theoretical predictions. Recent data (3.3-16) (not yet published), based on polyester support the m = exponent for thin fuel beds.

Velocity of Environment

The influence of forced convective motion on the flame spread rate is strong. While it has been observed that air flow in the direction of propagation increases the flame spread exponentially with flow velocity (3.3-17), this phenomenon is sufficiently complex to have discouraged extensive study. However, detailed studies of the influence of air flow opposed to a spreading flame that have been conducted (3.3-18, 19). These studies have shown not only that the increased spreading rate is a result of increased heat transfer ahead of the flame, but also that the spreading rate varies as

$$V \sim U\infty^{\bullet},$$

where V = spread rate, $U\infty$ = opposed flow velocity, $\bullet = 1/3$ for duct flow, and $\bullet = 1/2$ for free stream flow over a flat plate. Furthermore, at high flow velocities, a decrease in spreading rate was noted (3.3-18), which is probably due to an increase in the effects of chemical kinetic times on the spreading rate.

External Radiant Flux

As might be expected, augmentation of the heat transfer forward from the flame with an external radiant heat flux increases the spreading rate. The influence of external radiant flux on flame spread has been reported by Kwentus (3.3-20), Perkins and Pettett (3.3-21), Kashiwagi (3.3-22) and Hirano and Sato (3.3-5).

Environmental Composition

Both the oxygen mole fraction and inert diluent have been shown to affect the flame spread rate (3.3-2). As with many of the parameters already mentioned, the extent of this influence depends on whether or not the specimen is thermally thick or thermally thin. Generally,

$$V \sim (Y_{ox})^n ,$$

where V = spread rate, Y_{ox} = oxygen mole fraction of the surroundings, $n \approx 2$ for thermally thick specimens, and $n \approx 1$ for thermally thin specimens.

Summary of Flame Spread

A number of factors affect the flame spread rate in addition to the chemical composition of the fuel bed itself. Various theories have been proposed (3.3-10, 19, 23, 25) to rationalize some of these effects. Models differ widely on the assumptions made concerning the structure of the spreading flame and the mode(s) of heat transfer selected as being important. While all of these models have been successful in predicting some of the experimentally observed effects, each has been criticized for one reason or another. Sirignano reviewed some of the pertinent theories in 1972 (3.3-26). A current review is being prepared by F. A. Williams for the Sixteenth (International) Combustion Symposium.

In summary, while much has been learned about flame spread over solid fuel beds during the past decade, scientists still do not agree as to whether adequate methods exist for predicting the flame spread rate for a particular material, in a given geometric and environmental situation, even in the simplest cases.

3.3.4 Burning Rage of Plastic Slab

Once fully ignited, a thick plastic slab burns at a rate which is believed to be essentially independent of all chemical kinetic parameters, except those that might affect the luminosity of the flame. The theory relating this rate to the governing parameters is well advanced (3.4-1).

For the case of a small or nonluminous flame, radiation is negligible, and the burning rate is controlled by an energy balance at the surface, in which heat is convected from the flame gases to the surface and absorbed by conduction into the interior, endothermic heat-up, depolymerization, and gasification of the plastic (some of the heat is lost by re-radiation from the surface). While the depolymerization reactions obey the laws of chemical kinetics (i.e., proceed with finite activation energies), the kinetics have very little influence on the rate of gasification because the surface assumes a sufficiently high temperature to permit the reactions to occur just fast enough to satisfy the energy balance.

Furthermore, the gaseous flame reactions are describable by the Burke-Schumann diffusion flame model, or refinements thereof, in which the reaction rates are assumed to be infinite, and heat transfer as well as diffusion rates control the process.

For the case of laminar flow and negligible radiation, a numerical solution has been formulated. It agrees with experiment (3.4-2). The important parameters are the Spalding B-number (3.4-1), which is essentially the ratio of the heat of combustion to the heat of gasification times the stoichiometric fuel-air ratio, and the Grashof number, which is a ratio involving buoyancy force acting on an element of hot gas and viscous force. Weaker parameters are the surface temperature, the gas transport properties, and the stoichiometric ratio (in addition to its effect on the B-number).

When the specimen has dimensions of about 10 cm or more, the flame will be turbulent instead of laminar and large enough so that radiation from the flame to the surface becomes significant. In this case, neither the theory of turbulence nor the theory of flame radiation is sufficiently advanced to permit quantitative calculation of burning rate from first principles, as contrasted with the preceding laminar case. However, experimental studies of such turbulent flames have indicated the relative importance of radiation.

For example, a horizontal slab of polystyrene, 18 cm square, burning as a pool, is consumed nearly twice as fast (mass per unit area) as an identical slab of polyoxymethylene (3.4.3). In contrast, when a small specimen of polystyrene rod is burned in an apparatus with an opposed air jet (3.4-4), it is consumed at less than half as fast as polyoxymethylene. When burning as a large turbulent pool, polystyrene has a flame which is highly luminous, whereas a polyoxymethylene flame is blue. In small opposed-jet burners, both flames are blue.

Additionally, when a large polymethyl methacrylate slab is burned vertically (on one side) instead of horizontally, the burning rate is found to be slower. An explanation for this phenomenon is that a point on the surface of the vertical slab doesn't "see" as great a thickness of flame as the corresponding point on the horizontal slab, because the flame shape is different and, accordingly, less radiation is received. It has been possible to make direct measurements of the radiant and convective heat arriving at the surface of a vertical polymethyl methacrylate slab; at a point 50 cm from the bottom, the incident radiant flux is 2.6 times as large as the convective flux from the flame (3.4-5). Thus, even larger radiant feedback would be expected from the horizontal burning configuration.

Furthermore, the rate of increase in burning rate for burning plastics, with incident radiant energy supplied by electrical radiant panels, has been measured (3.5-3). For example, a radiant flux of 1.3 watts/cm^2 increased the steady burning rate of a vertical polymethyl methacrylate slab by a factor of 2.5. A ceiling of unit emissivity would emit such a radiant flux if it reached a temperature of 686 °K. Large fires emit radiant fluxes at least five times greater than this value. Clearly, this effect can be of dominant importance.

Burning rates of thick slabs of wood or charring plastics are much lower than those of melting plastics, for two main reasons: 1) the char layer acts as insulation between the flame and the virgin fuel; 2) since the surface temperature (T_s) is about 900 °K instead of about 650 °K, the re-radiation of heat from the surface, which varies as T_s^4, is much higher.

These recently acquired insights into the role of radiation in burning, when coupled with better knowledge of radiant emission and transmission characteristics of flames, as well as the role of turbulence in mass and energy transport processes, give promise for a better understanding of these phenomena.

3.3.5 Burning in an Enclosure

A wood crib in the open burns at a reproducible rate, which may be correlated with the geometrical parameters of the crib (3.5-1). When such a crib is burned in a noncombustible enclosure with minimal ventilation, the burning rate is greatly reduced, and the proportion of carbon monoxide in the combustion products is greatly increased (3.5-2).

However, if the test is repeated in an enclosure that is very well ventilated, the burning rate is considerably higher (up to 70% more) than the burning rate for the same crib in the open (3.5-3). The probable explanation for this effect is that the ceiling above the crib gets hot and radiates the heat back to the crib. It is also possible that the hot smoky gases layered under the ceiling also radiate significant heat to the crib. While the relative radiative effects of the ceiling and the hot gases under the ceiling have not yet been differentiated, it is clear that the radiant feedback is important.

In a more dramatic experiment, a horizontal slab of polymethyl methacrylate, 30 inches (71.2 cm) square, was burned under a ceiling 4 feet (1.2 m) high, with adequate ventilation. The slab burned for a short initial period at a rate corresponding to burning in the open, but then rapidly accelerated to a rate about four times a sgreat, as radiant feedback built up (3.5-3). Kwagoe (3.5-4) established that the burning rate of a fully developed fire in an enclosure with restricted ventilation is proportional to $Ah^{1/2}$, where A is the area and h is the height of the window opening. However, the limits to conditions for which this law applies are not fully understood. Since the burning of materials in compartments is the principal way of studying smoke or toxic gas production, further development of the laws governing such burning is essential.

3.3.6 Movement of Smoke

To date, only limited ability exists to predict the rate at which smoke and combustion products move from a flaming or smoldering region to the more remote parts of a building or structure, or the rate of dilution as this movement occurs. However, research is now underway to develop scientific tools for such prediction. The relevance of such research to smoke detector location and personnel escape potential should be obvious.

Emmons (3.6-1), at Harvard, is developing a means for calculating the bouyantly driven flow of fire gases through a window or door of a burning room. He assumes a fire of given intensity and attempts to calculate the inflow of air, the outflow of products, and the location of the interface between the two. Simple stoichiometric calculations do not suffice, because: a) mixing occurs between the two flows b) secondary flows are induced by wall-cooling. Emmons attempts to treat the problem by a hydraulic analogy in which the flow of hot gases under the top of the opening is an inverted analogue of water flowing over a weir. An empirical coefficient must be introduced to obtain an accurate answer. Further work is required.

Zukoski (3.6-2), at California Institute of Technology, is concerned with the rate of rise of smoke and fire gases up an open stairwell. He is studying this flow with an analogy using dyed water rising through denser salt water in a vertical tube. His goal is to develop dimensionless parameters describing the movement.

Alpert (3.6-3), at Factory Mutual, has developed and confirmed a theory for predicting the motion and dilution rate of turbulent fire gases impinging on an "infinite" flat ceiling and then moving radially outward as a "ceiling jet." For a given fire intensity and ceiling height, the distributions of velocity, temperature, and smoke concentration can be predicted at any point. Of course, the smoke generation rate must be separately known. The analysis can also be applied to a fire in a corner or adjacent to a plane wall.

Whereas steady-state flow under a flat ceiling treated by Alpert is tractable by analytical techniques, the more general case of transient flow following onset of fire, and a complex ceiling geometries, is considerably more difficult. However, a partial solution to this problem exists in a modeling theory developed by Heskestad (3.6-4). According to this theory, the transient environment actually measured for one set of conditions (ceiling height, fire growth rate) can be used to predict the transient environment for other conditions.

A complete analysis of transient fire gas motions in a closed "room" with cylindrical symmetry has been performed by Torrance and Rockett (3.6-5) utilizing a computer. In their analysis, a "hot spot" of arbitrary temperature and size suddenly appears on the center of the floor, and resulting bouyant motions are obtained. It was necessary, however, to make the unnatural assumption of laminar flow. If this approach could be extended into the turbulent flow regime, and asymmetries such as a window could be introduced, this approach would become promising.

Clearly, dynamic analysis of bouyant motion of smoke and fire gas is in its infancy. Should this tool be properly developed, it would permit analyses of successively more complex geometries, and finally, real fire situations. As long as a fully rational method of analysis is lacking, there will be no reliable way to generalize a small number of large-scale fire test results.

3.3.7 Extinguishment

An important complementary aspect of the fire hazard associated with a given polymeric material is, once the material has ignited and begins to burn, how easily can the fire be extinguished. The answer to this question depends on the extinguishment approach employed as well as on the fire conditions that prevail.

Various approaches to solving the problem of extinguishment unwanted fires are as follows:

1. isolation of the fuel
2. isolation of the oxygen available to the fire
3. cooling and solid fuel
4. cooling or diluting the gas phase
5. inhibition of the chemical reaction homogeneously
6. inhibition of the chemical reaction heterogeneously

The extinguishment of fires by water has long been a common practice. Since water is usually readily available, inexpensive, and generally effective, many industrial properties and public buildings are protected by automatic sprinkler systems that employ water as the extinguishing agent. Water, being nontoxic and having a very high heat absorbing capacity, will continue to be employed as the suppressing agent in many fire situations.

Some investigations of extinction of fires by water, particularly those that dealt with pool fires and wood crib fires (3.7-1, 2, 3, 4) have focused on several facets of the problem, such as: 1) the extent of penetration of the water drops through the fire plume to the burning fuel surface, 2) the drop size distribution of commonly available standard sprinkler heads and the interacion of these drops with bouyant fire plumes, and 3) the effects of burning rate, pre-burn time, water droplet weight and mean diameter, as well as water application rate on extinction. Results of these studies indicate that: 1) only a fraction of the drops produced by a standard sprinkler actually reach the burning surface; 2) the larger the droplet diameter, the more effective it is in penetrating the fire; and 3) the burning rate, pre-burn time, water droplet weight and mean diameter, and water application rate are all important to the extinction process.

A recent study by Magee and Reitz (3.7-5) on the extinguishment of radiation-augmented plastic fires by water spray focused on 1) the effectiveness of water in suppressing the fire, and 2) the critical water application rates required for extinction as a function of radiant flux and type of plastic. Results of this study indicate that the role of water in suppressing a fire is principally to cool the fuel surface. For example, the ratio of the cooling effect of water at extinguishment to the total heat released by the fire is typically 3-5 percent. On the other hand, the cooling effect of water is from 0.5 to 1.1 times the net energy being transferred to the solid by the flame. This finding is supported in practice where firefighters are usually instructed to apply the water to the condensed fuel, not to the flame. This study also determined the critical water application rate ($gm/cm^2/s$) to the fuel surface necessary to extinguish plastic fires to varying intensities. For cellulosic fires, Rasbash (3.7-1) has suggested, as a rough criterion, that reducing the flame temperature below 1580 °K, or extracting from the condensed fuel 2500 calories per gram of fuel gasified, will reduce extinction. Magee and Reitz (3.7-5) found that for plastic fires, removal at the surface of from 170 to 450 calories per gram of fuel gasified produced extinction.

Water, steam and fogging nozzles are also designed to achieve fire suppression by cooling the hot combustion gases. One study reported that, under certain conditions, this approach is less efficient than cooling the condensed phase directly (3.7-1). Further study of minimum water requirements for both application modes is needed under various fire conditions.

Carbon dioxide extinguishers function primarily by diluting the oxidizer, or isolating the oxidizer from the fuel; however, modest benefits may accrue from some cooling of the gas phase and condensed phase as well. Frequently, NFPA regulations specify the size and number of CO_2 extinguishers required in various structures, yet data are absent on the amount of extinguishing agent and the application rate necessary to suppress different types of fires.

Homogeneous chemical inhibitors, for example, halogen-containing compounds such as CF_3Br (Halon 1301), have been found effective in suppressing hydrocarbon fires, as well as class A fires, as long as they are not deep-seated. The effectiveness of these agents and the suppression mechanisms involved have been the subject of extensive study (3.7-6, 7, 8, 9, 10). The mechanisms of suppression are still the subject of much debate. Moremover, information regarding the dependence of inhibition requirements on various fire parameters, fuel type, fire intensity, etc., is lacking.

Alkali metal salts, such as sodium and potassium bicarbonate, are the principal heterogeneous chemical inhibitors. Effects of therse inhibitors on fire suppression have also been reviewed (3.7-6, 7, 8, 9). A recent study (3.7-10) seems to have established that their inhibiting mechanism is not heterogeneous (occurring either on the surfaces of the alkali metal salt particles or on the surfaces of condensed oxide particles produced through chain breaking of oxidation reactions by alkali metal hydroxides. Further support of a homogeneous inhibition mechanism for potassium salts has been offered (3.7-11).

3.4 Applications of Fire Dynamics to Test Method Development

In general, our knowledge of fire dynamics is still primitive; it is difficult to point to many examples of its use in test method development. Also, most tests were developed when such knowledge was even more primitive, or at times when enormous pressure existed to develop test methods in a matter of months, permitting no time for a realistic fire dynamics approach. Even now tests are often introduced in response to a specific disaster rather than as part of a long-range development plan.

However, in a recent case, fire dynamics considerations were used in test development for fire spread down a corridor with flammable floor covering. In this case, corridor fire experiments were performed, and the radiation from the ceiling, as well as the burning gases just below the ceiling, were identified as critical in promoting a flameover condition'' (propagation rates of the order of a foot per second) over the floor covering. thus, a test is being developed involving radiant heat impinging on a floor-covering sample, with piloted ignition at one end (see Figure 2).

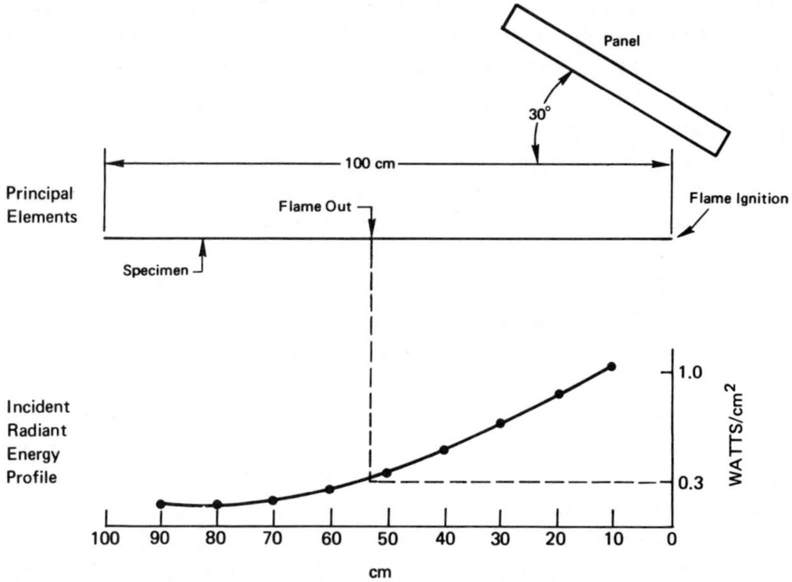

Figure 2. The flooring radiant panel test

In the Flooring Radiant Panel Test, the flame starts in a region of high radiant flux and propagates to regions of progressively lower radiant flux; then, at some point, the flame ceases to advance. The acceptance must not occur. This critical flux, of the order of 0.25 to 0.65 watts/sq cm, would be set depending on the occupancy considered. These fire dynamics studies (3.8-1, 2, 3, 4, 5) established the importance of this variable and provided guidance as to a proper value for the critical flux.

Another approach to test method development is the use of fire modeling. Some historical background on this subject was presented in a 159 symposium (3.4-1, 3.8-6, 7, 8, 9). In the fire modeling approach, a physical model of a fire situation is reduced in scale while maintaining geometric similarity and preserving the important chemical and thermo-dynamic properties of the materials involved. Were such an approach successful, the benefits to fire testing would be obvious. Unfortunately, progress in its development has been very slow.

The difficulties of modeling a fire by reducing scale arise in several ways: 1) very small fires are laminar while larger fires are turbulent; 2) as far as fluid mechanics is concerned, the ratio of bouyancy forces to viscous forces in the convective flow of fire gases is size-dependent; 3) the radiant emission and self-absorption of the flame are size-dependent; 4) the gas-phase time scale in the fire is shorter for small than for large fires, with possible effect on incomplete combustion products. These formidable difficulties have kept knowledgeable people from having much confidence in fire model test results.

However, it is anticipated that valid modeling procedures may be developed for at least some aspects of fire behavior as the understanding of fire dynamics relevant to situations of interest improves. Then, compensation for any errors or distortions introduced by the modeling might be achieved by varying other parameters such as ambient temperature, pressure, oxygen concentration, or ambient radiation. For example, it has been suggested that the gravitational force may be varied by use of a centrifuge.

Some progress has been made by a pressure-modeling technique (3.8-6), which is based on the principle that the Grashof number is invariant if the product of pressure squared and size cubed is held constant. The Grashof number is a dimensionless ratio of buoyancy forces to viscous forces in the hot gases around a fire. Convective heat transfer from the fire plume to the surroundings, expressed as a dimensionless quantity, is governed by the Grashof number. Pressure-modeling has been shown to be valid for diffusion-controlled burning as long as radiation either varies with pressure in certain prescribed ways or is negligible. It has worked over a tenfold specimen size range for turbulent burning of polymethyl methacrylate and wood objects in several shapes. It is also possible to model the temperature and velocity distributions near a fire in a building, by using a reduced-scale physical model at atmospheric pressure (if the fire intensity is controllable) and reducing the fire intensity so as to maintain a constant Froude number (the ratio of the square of the maximum fire plume velocity to the ceiling height) — see de Ris (3.8-7). This modeling procedure appears to be valid except in the viscous boundary layer. Future progress in the modeling of fire situations will be difficult to achieve without a better understanding of the underlying controlling mechanisms of fire.

3.5 Conclusions

1. Fire dynamics is emerging as a scientific discipline. It encompasses chemistry, physics, mathematics, fluid mechanics and heat and mass transfer.

2. The complex dynamic interaction of fundamental processes in fire dynamics coupled with a relatively small funding effort, has led to rather slow progress in quantifying fire behavior.

3. fire dynamics funding for FY 1977 is estimated to *decrease* 26% below the FY 1976 level, because of federal government budgetary actions. An *increase* of perhaps 8% is needed even to maintain the past level of effort, because of inflation.

4. Our current level of understanding of fire dynamics is not sufficiently developed to permit thorough scientific analysis of most fire test methods, fire scenarios or fire modeling procedures.

5. The small amount of relevant scientific research being supported, chiefly in the United States, England and Japan, is continually leading to a better understanding of

important aspects of fire behavior. Thus, while a certain body of fire dynamics knowledge exists, it is in a highly incomplete state, and many years of intensive research are needed to fill in the gaps.

6. The behavior of a given material in a fire is dependent not only on the properties of the fuel, but also on the fire environment to which it is exposed. Consequently, if test methods are to be meaningful, they must simulate the critical fire dynamic conditions. An understanding of fire dynamics is essential, if the *critical* conditions are to be identified. Consequently fire dynamics can be extremely useful in test method selection and development.

7. Valid modeling procedures, based on evolving fire dynamic principles, offers promise for reduction in the dependence on, and the cost of, large scale fire tests.

3.6 Recommendations

To obtain the full benefits of the scientific approach to improved fire safety, it is essential to implement recommendations as follows:

1. An increased program of sufficient magnitude and stability, funded by the government, in the specific area of fire dynamics, must be established.

2. This program should be conducted primarily in academic nonprofit, and governmental research organizations, but the projects should be closely monitored by advisory boards including broad representation by manufacturers, users, and members of standards-setting organizations.

3. At least an additional 50 scientists and engineers (approximately equal to the current effort) should be supported over a 10-year period in the specific area of fire dynamics research. This increase would represent an additional cost of approximately $\$2.5 \times 10^6$ per year; approximately 0.02% of the annual cost to society of fire in the U.S. (President's Commission on Fire Protection and Control estimate for 1972 was some $11.4 billion).

4. In view of the shortage of qualified engineers and scientists in the appropriate disciplines, the program should be brought to full strength during the first four years.

5. Fire dynamic principles should be taught to fire protection engineers, code officials, architects, and designers.

6. Fire dynamics expertise should be employed when developing new test methods and for validation and improvement of existing test methods.

3.7 References

3.3-1 Lewis, B. and von Elbe, G., "Combustion, Flames, and Explosions of Gases," Academic Press, New York, 1961.

3.1-2 Williams, F. A., "Combustion Theory," Addison-Wesley, Reading, Massachusetts, 1965.

3.1-3 N.A.S-N.R.C. Publication 786, "The Use of Models in Fire Research." Washington, D. C., 1961.

3.1-4 N.B.S. Special Publication 357, "The Mechanisms of Pyrolysis, Oxidation, and Burning of Organic Materials," G. P. O., Washington, D. C., 1972.

3.2-1 Price, E. W., Bradley, H. H., Jr., Dehority, G. L. and Ibirim, M. M., "Theory of Ignition of Solid Propellants," AIAA J., *44*, pp. 1158–1181, (1966).

3.2-2 Frank-Kamenetskii, D. A., "Diffusion and Heat Exchange in Chemical Kinetics," Princeton University Press, Princeton, N. J., (1955), translation.

3.2-3 Hicks, B. L., "Theory of Ignition Considered as a Thermal Reaction," J. Chem. Phys., *22*, ppg. 441–429, (1954).

3.2-4 Thomas, P. H., "Self-Heating and Thermal Ignition – A Guide to its Theory and Application," ASTM Special Tech. Publ. 502, Philadelphia, pp. 56–82, (1972).

3.2-5 Merzhanov, A. G. and Averson, A. E., "Present State of the Thermal Ignition Theory." Combustion and Flame, *16*, pp. 89–124, (1972).

3.2-6 Shouman, A. R., Donaldson, A. B. and Tsao, H. Y., "Exact Solutions to the One-Dimensional Stationary Energy Equation for Self-Heating Solid," Combustion and Flame, 23, pp. 17–28, (1974).

3.2-7 Thomas, P. H., "An Approximate Theory of 'Hot Spot' Critically," Combustion and Flame, *21*, ppg. 99–109, (1973).

3.2-8 Zaturska, M. B., "Thermal Explosion of Interacting Hot Spots," Combustion and Flame, *25*, pp. 25–30, (1975).

3.2-9 Welker, J. R., "The Pyrolysis and Ignition of Cellulosic Materials: A Literature Review,: J. of Fire and Flammability, *1*, pp., 12–29, (1970).

3.2-10 McAlevy, R. F., III, Cowan, P. L. and Summerfield, M., "The Mechanism of Ignition of Composite Propellants by Hot Gases," ARS Preprint 1058–60, (Janury 1960); also ARS Progress in Astronautics and Rocketry: Solid Propellant Rocket Research, edited by M. Summerfield, (Academic Press, Inc., New York, 1960), *1*, pp. 623–652.

3.2-11 Hermance, C. R., Shinnar, R. and Summerfield, M., "Ignition of an Evaporating Fuel in a Hot Oxidizing Gas, Including the Effect of Heat Feedback," Astro. Aeta, *12*, pp. 94–112, (1966).

3.2-12 Hermance, C. E., "Ignition of Polymeric Materials Stimulated by Rapid External Heating." Paper presented at the Flammability Characteristics of Polymeric Materials Conference, Polymer Conference Series, University of Utah, June 21–26, 1971.

3.2-13 Ohlemiller, T. J. and Summerfield, M., "Radiative Ignition of Polymeric Materials on Oxygen/Nitrogen Mixtures," Thirteenth Symposium (International) on Combustion, The Combustion Institute, Pittsburgh, Pa., p. 1087, (1971).

3.2-14 Kashiwagi, T., "A Radiative Ignition Model of a Solid Fuel," Combustion Science and Technology, *8*, pp. 225–236, (1974).

3.2-15 Rangaprasad, N., Sliepcevich, C. M. and Welker, J. R., "The Piloted Ignition of Cotton Fabrics," J. of Fire and Flammability, *5*, pp. 107–115, (1974).

3.2-16 Simms, D. L., "Ignition of Cellulosic Materials by Radiation," Combustion and Flame, *4*, pp. 293–300, (1960).

3.2-17 Martin, S., "Diffusion Controlled Ignition of Cellulosic Materials by Intense Radiant Energy." Tenth Symposium (International) on Combustion, The Combustion Institute, Pittsburgh, Pa., pp. 877–896, (1965).

3.2-18 Akita, K., "Studies on the Mechanism of Ignition of Wood," Report of Fire Research Institute of Japan, N. 9, 1, 51, 97, 99, (1959).

3.2-19 Kashiwagi, T., MacDonald B. W., Isoda, H. and Summerfield, M., "Ignition of a Solid Polymeric Fuel in a Hot Oxidizing Gas Stream." Thirteenth Symposium (International) on Combustion. The Combustion Institute, Pittsburgh, Pa., pp. 1073–1086, (1971).

3.3-1 Friedman, R.; "A Survey of Knowledge About Idealized Fire Spread Over Surfaces," Fire Research Abstracts and Reviews, *10*, pp. 1–8, (1968).

3.3-2 Magee, R. S. and McAlvey, R. F., III, "The Mechanism of Flame Spread," J. Fire and Flammability, *2*, pp. 271–296, (1971).

3.3-3 Hirano, T., Noreikis, S. E. and Waterman, T. E., "Postulations of Flame Spread Mechanisms," Combustion and Flame, *22*, pp. 353–363, (1974).

3.3-4 Hirano, T., Norekis, S. E. and Waterman, T. E., "Measured Velocity and Temperature Profiles Near Flames Spreading Over a Thin Combustible Solid," Combustion and Flame, *23*, pp. 83–96, (1974).

3.3-5 Hirano, T. and Sato, K., "Effects of Radiation and Convection on Gas Velocity and Temperature Profiles of Flames Spreading Over paper." Fifteenth Symposium (International) on Combustion, The Combustion Institute, Pittsburgh, Pa., pp. 233–242, (1975).

3.3-6 Sibulkin, Mr., Ketelhut, W. and Feldman, S., "Effect of Orientation and External Flow Velocity on Flame Spreading Over Thermally Thin Paper Strips," Combustion Science and Technology, *9*, pp. 75–77, (1974).

3.3-7 Sibulkin, M. and Lee, C. K. "Flame Propagation Measurements and Energy Feedback Analysis for Burning Cylinders," Combustion Science and Technology, *9*, pp. 137–148, (1974).

3.3-8 Sibulkin, M. and Hansen, A. G., "Experimental Study of Flame Spreading Over a Horizontal Fuel Surface," Combustion Science and Technology, *10*, pp. 85–92, (1975).

3.3-9 Moussa, N. A. Toong, T. Y. and Backer, S., "An Experimental Investigation of Flame-Spreading Mechanisms Over Textile Materials," Combustion Science and Technology, *8*, pp. 167–175, (1973).

3.3-10 Fernandez-Pello, A. and Williams, F. A., "Laminar Flame Spread Over PMMA Surfaces," Fifteenth Symposium (International) on Combustion, The Combustion Institute, Pittsburgh, Pa., pp. 217–231, (1975).

3.3-11 Hansen, A. and Sibulkin, M., "Flame Spreading From and Point Source of Ignition on a Vertical Fuel Surface," Combustion Science and Technology, *9*, pp. 173–176, (1974).

3.3-12 Orloff, L., de Ris, J. and Markstein, G. H., "Upward Turbulent Fire Spread on Burning of Fuel Surface," Fifteenth Symposium (International) on Combustion, The Combustion Institute, Pittsburgh, Pa., pp. 183–192, (1975).

3.3-13 Orloff, L., Private Communication, (1975).

3.3-14 Markstein, G. H. and de Ris, J., "Upward Fire Spread Over Textiles," Fourteenth Symposium (International) on Combustion, The Combustion Institute, Pittsburgh, Pa., pp. 1085–1097, (1973).

3.3-15 Frey, A. E. and T'ien, J. S., "Near Limit Flame Spread Over Cellulosic Materials." Paper presented at the 1974 Fall Meeting, Eastern Section/The Combustion Institute, Applied Physics Laboratory, Silver Spring, Maryland, November 12–13, 1974.

3.3-16 Granzow, A. Private Communication, American Cyanamid Company, Bound Brook, New Jersey, (1975).

3.3-17 Rothermel, R. C. and Anderson, H. E., "Fire Spread Characteristics Determined in the Laboratory," U.S. Forest Service, Research Paper INT-30, 1966, Inter-mountain forest & Range Experiment Station, Ogden, Utah.

3.3-18 Pagni, P. J. Private Communication, University of California, Berkeley, California, (1975).

3.3-19 Lastrina, F. A., Magee, R. S. and McAlevy, R. R., III, "Flame Spread Over Fuel Beds: Solid Phase Energy Considerations." Thirteenth Symposium (International) on Combustion, The Combustion Institute, pp. 935–949, (1971).

3.3-20 Kwentus, G. K., "Fuel Preheating in Free-Burning Fires," Ph. D. Thesis, Massachusetts Institute of Technology, (1967).

3.3-21 perrins, L. E. and Petlett, K., "Measurement of Flame Spread Velocities," J. Fire and Flammability, *5*, pp. 85–102, (1974).

3.3-22 kashiwagi, T., "A Study of Flame Spread Over a Porous Material Under External Radiative Fluxes," Fifteenth Symposium (International) on Combustion, The Combustion Institute, Pittsburgh, Pa., pp. 255–265, (1975).

3.3-23 de Ris, J., "The Spread of a Laminar Diffusion flame," Twelfth Symposium (International) on Combustion, The Combustion Institute, Pittsburgh, Pa., pp. 241–252, (1969).

3.3–24 Tarifa, C. S., Notario, P. P. and Torralbo, A. M., "On the Process of Flame Spreading Over the Surface of Plastic Fuels in an Oxidizing Atmosphere," Twelfth Symposium (International) on Combustion, The Combustion Institute, Pittsburgh, Pa. pp. 229–240, (1969).

3.3-25 Sirignano, W. A., "A Critical Discussion of Theories of Flame spread Across Solid and Liquid Fuels," Combustion Science and Technology, *6*, pp. 95–105, (1972).

3.3-26 Backer, S., Tesoro, G. C., Toong, T. Y. and Moussa, N. A., "Textile Fabric Flammability." MIT Press, 1976.

3.4-1 Friedman, R., "Aerothermodynamics and Modeling Techniques for Prediction of Plastic Burning Rates," J. of Fire and Flammability, *2*, pp. 240–256, (1971).

3.4-2 Kim, J. S., de Ris, J. and Kroesser, F. W., "Laminar Free-Convective Burning of Fuel Surfaces," Thirteenth Symposium (International) on Combustion, Combustion Institute, pp. 949–961, (1971).

3.4-3 Magee, R. S. and Reitz, R. D., "Extinguishment of Radiation-Augmented Plastic Fires by Water Sprays," Fifteenth Symposium (International) on Combustion, The Combustion Institute, pp. 337–347, (1975).

3.4-4 Holve, D. J. and Sawyer, R. F., "Diffusion Controlled Combustion of Polymers," Fifteenth Symposium (International) on Combustion, The Combustion Institute, pp. 351–361, (1975).

3.4-5 Orloff, L., de Ris, J. and Markstein, G. H., "Upward Turbulent Fire Spread and Burning of Fuel Surfaces," Fifteenth Symposium (International) on Combustion, The Combustion Institute, pp. 183–192, (1975).

3.5-1 Block, J. A., "Nonpropagating Free-Burning Fires," Thirteenth Symposium (International) on Combustion, The Combustion Institute, pp. 971–978, (1971).

3.5-2 Tewarson, A., ,,Experimental Fires in Enclosures: I. Cellulosic Materials," Combustion and Flame, *19*, pp. 101–111, (1972).

3.5-3 Alpert, R. and Croce, P., in Semi-Annual Progress Report, The Home Fire Project, NSF-RANN, Harvard University – Factory Mutual Research, December 1973.

3.5-4 Heskestad, G., "Modeling of Enclosure Fires," Fourteenth Symposium (International) on Combustion, The Combustion Institute, pp. 1021–1030, (1973).

3.6-1 Emmons, H. W., "National Convective Flow Through an Opening." Harvard University, Cambridge, Massachusetts. N.S.F. Grant GI 30957, Technical Report No. 1, 1974. To be published.

3.6-2 Zukoski, E. E., "Convective Flows of Building Fires." California Institute of Technology, Pasadena. Work in progress under N.S.F. Grant, 1974.

3.6-3 Alpert, R. L., "Turbulent Ceiling-Jet Induced by Large-Scale Fires." Factory Mutual Research Corp., Norwood, Massachusetts. Combustion Science and Technology, 197–213, (1975).

3.6-4 Heskestad, G., "Similarity Relations for the Initial Convective Flow Generated by Fire." ASME Paper No. 72-WA/HT-17, 1972.

3.6-5 Torrance, K. E. and Rockett, J. A., "Numerical Study of Natural Convection in an Enclosure with Localized Heating From Below – Creeping Flow to the Onset of Laminar Instability," J. Fluid Mechanics, *36*, Part 1, pp. 33–54, (1969).

3.7-1 Rasbash, D. J., "The Extinction of Fires by Water Sprays," Fire Research Abstracts and Reviews, *4*, January–May 1962, pp. 28–53.

3.2-2 Byran, J., Engineering, 159, 457, 1945.

3.7-3 yao, C. and Kalelkar, A. S., "Effect of Drop Size on Sprinkler Performance, Fire Technology, *6*, p. 254, (1970).

3.7-4 Kalelker, A. S., "Understanding Sprinkler Performance: Modeling of Combustion and Extinction." Paper presented at the 75th Annual Meeting National Fire Protection Association, San Francisco, California, May 1971.

3.7-5 Magee, R. S. and Reitz, R. D., "Extinguishment of Radiation Augmented Plastic Fires by Water Sprays," Fifteenth Symposium (International) on Combustion, The Combustion Institute, Pittsburgh, Pa., pp. 337–347, (1975).

3.7-6 Friedman, R. and Levy, J. B., "Survey of Fundamental Knowledge of Mechanisms of Action of Flame Extinguishing Agents." WADC Technical Report 56-568, U.S. Department of Commerce, (1957).

3.7-7 Friedman, R. and Levy, J. B., "Mechanism of Action of Chemical Agents for Flame Extinguishment," Fire Research Abstracts and Reviews, *1*, pp. 81–88, (1959).

3.7-8 Fristrom, R. M., "Combustion Suppression," Fire Research Abstracts and Reviews, *9*, pp. 125–160, (1967).

3.7-9 McHale, E. T., "Survey of Vapor Phase Chemical Agents for Combustion Suppression," Fire Research Abstracts and Reviews, *11*, pp. 90–104, (1969).

3.7-10 Iya, K. S., Wollowitz, s. and Kaskan, W. E., "The Mechanism of Flame Inhibition by Sodium Salts," Fifteenth Symposium (International) on Combustion, The Combustion Institute, pp. 329–336, (1975).

3.7-11 McHale, E. T., "Flame Inhibition by Potassium Compounds," Combustion and Flame, *24*, pp. 277–279, (1975).

3.8-1 Hartzell, L. G., "Development of a Radiant Panel Test for Flooring Materials," J. Fire and Flammability/Consumer Prod. Flammability *1*, pp. 376–389, (1974).

3.8-2 Denyes, W. and Quintiere, J., "Experimental and Analytical Studies of Floor Covering Flammability With a Model Corridor." National Bureau of Standards NBSIR 73-199, May 1973. 105 pp., and Quintiere, J. G., "A Characterization and Analysis of NBS Corridor and Performance of Floor Covering Materials." National Bureau of Standards NBSIR 75-69, June 1975, 83 p.

3.8-3 Kashiwagi, T., "A Study of Flame Spread Over a Porous Material Under External Radiation Fluxes," Fifteenth Symposium (International) on Combustion, The Combustion Institute, pp. 255–265, (1975).

3.8-4 Quintiere, J., "Some Observations of Building Corridor Fires," Fifteenth Symposium (International) on Combustion, The Combustion Institute, Pittsburgh, Pa., pp. 163–174, (1975).

3.8-5 Benjamin, I. A. and Adams, C. H., "The Flooring Radiant Panel Test and Proposed Criteria," Fire Journal, *70*, pp. 63, (1976).

3.8-6 Berl, W. G. (Editor), "The Use of Models in Fire Research." National Academy of Sciences–NRC., Publication 786, Washington, D. C., (1961).

3.8-7 de Ris, J., Kanury, A. M. and Yuen, M. L., "Pressure Modeling of Fires." Fourtheenth Symposium (International) on Combustion, The Combustion Institute, pp. 1033–1044, (1973).

3.8-8 de Ris, J., "Modeling Techniques for Predicting Fires." Applied Polymer Symposium No. 22, J. Wiley, PP. 185–193, (1973).

3.8-9 Emmons, H., "Heat Transfer in Fire." Journal of Heat Transfer, May 1973, pp. 145–151.

---

[1]During a transitional period FY 75–77, NFPCA (National Fire Protection and Control Administration) has directly funded a substantial portion of fire dynamics research.

[2]References start on page 41.

[3]J.R. Brown and P. Dunn, Report 561, Australian Defence Sci. Service, Defence Stds. Labs. Moribyrnong, Victoria, Australia, June 1973.

CHAPTER 4

THE FIRE SCENARIO METHOD: THE ESSENTIAL APPROACH TO IMPROVED FIRE SAFETY

4.1 Introduction

A prominent window poster set forth an admonition as follows:

1,000,000 buildings will burn
300,000 people will be injured
12,000 people will die
Learn how to prevent fires
National Fire Prevention Week

It was read by many, but heeded by few. In the ensuing year, its predictions came true. Much more than is currently being done is required to turn the trend in fire damage around.

The objective to this chapter is to consider the problem of fire prevention in a broad context. Evidence demonstrates that, in addition to polymeric materials, people are involved, directly or indirectly. Decisions people make and their actions often determine whether a fire will or will not occur. Interviews with major manufacturers and others concerned with fire hazards demonstrate that most individuals faul to recognize their role in creating a fire hazard.

Growth in the use of disposable packaging and other utility items in the home, as well as the proliferation of unsolicited mail, might be considered a case in point under the heading of "Trash", which frequently accumulates in unattended areas of the home to create fire hazards. No particular individual, grocery store, purveyor of paper packaged goods, or mailing list house can be brought to task for similar hazards, yet they exist.

A Swiss meat fondue pot, heated by a can of solidified fuel, provides another example of serious risk. By itself, when probably used, it is not hazardous, but, when placed on a structural foam table covered with a disposable table cloth, laden with disposable cups, napkins and plates, it waits to ignite. None of the suppliers or manufacturers deliberately set out to turn a party into a holocaust, but that's what happened when a housewife attempted to create a festive mood for a special family event. A guest tipped the fondue pot; what followed next was predictable. Hot oil soaked the paper. The flame from the "canned heat" ignited the oil soaked paper, which in turn ignited the structural foam. There was no time for people to react effectively. A feeble attempt was made to smother the rapidly developing fire. By the time the fire department was called on a neighbor's telephone and arrived, the fire engulfed the rest of the house.

Another example involves a thermally activated elevator call button and photoelectric doorstep that gave store patrons and passengers a feeling of security. All was well until a fire broke out on one of the floors. Heat summoned one of the elevators, which was loaded with people. The doors opened and smoke from the fire locked it open. The occupants of the elevator became victims of the fire. While such devices were developed by a combination of highly skilled professionals, they did not consider the effects of a fire on device operation.

These examples illustrate not only what can happen, but also what is happening. Industries want to provide society with greater conveniences at less cost, but many products, harmless in themselves, can be combined to form hazards.

Accordingly, the committee believes fire prevention cannot be achieved by an attack on materials alone. Moreover, as a nation we are poorly prepared to cope with fire. A new system of communication is required to make people aware of the fact that they *do* play an important role in fire prevention. In this chapter the Fire Scenario Method for hazard identification is recommended as an integrated approach to improved fire safety.

4.2 The Fire Scenario

A fire scenario is a generalized, detailed description of a fire incident. When properly developed, a scenario describes all essential elements of a fire incident, e.g., the ignition source, the first material ignited, how the fire spread, the time to flashover, how it was contained and finally extinguished.

Several kinds of fire scenarios exist. The first type is a strictly factual account of a specific fire incident. Generally, this type of scenario is impossible to generate accurately since it is extremely difficult to obtain qualitative data on all aspects of a real fire. A second type is the hypothetical fire scenario in which fact and supposition are combined to develop a complete description of a specific fire; or, elements of several different fires are synthesized into a single plausible scenario. The third type is a theoretical fire scenario; a paper study in which one models the growth of a fire, in a real situation, from fire dynamic principles. The fourth type is an experimental fire scenario in which a fire is deliberately started in a pre-arranged setting to permit fire growth to be monitored and recorded.

4.3 The Fire Scenario Method

The Fire Scenario Method involves the generation and analysis of fire scenarios. Its major objective is to pin-point hazards and suggest means for fire prevention and control. It is analogous to the Case Study Method developed by the Harvard Business School as an aid for students attempting to sort out the multifaceted problems encountered in the business world. In the Case Study Method, each case represents an actual and factual account of real people in a real business situation. It gives no hint as to what the characters had done correctly or incorrectly or what they could have done better (although the intent of the case writer is contained in notes for the instructor.) Having discovered a problem, the student is expected to develop more attractive alternative courses of action. The Case Study Method has proved to be an excellent mediuim for the study of complex issues on a wide variety of topics.

Since the Fire Scenario Method involves an in-depth analysis of a specific fire scenario, which is not necessarily a factual account of a specific fire incident, it is important that the scenarios generated be realistic and contain the essential fire dynamic elements. Guidelines for developing fire scenarios are outlined in Chapter 5.

Once a scenario has been generated, persons concerned with improved fire safety have pertinent aspects before them in the form of a case study and the situation may be examined in terms of future prevention and control. For example, could any of the participants in a fire have been better instructed or trained? Could more fire safe materials have been employed? Or, would a design modification have eliminated or controlled the hazard? Various alternative courses of action to improve fire safety could emerge from such analyses, and these alternatives can then be considered in the context of other constraints and/or objectives. Unfortunately, the committee found little evidence that the case study approach has been used to improve fire safety.

4.4 Use of Fire Scenarios in Education

The committee proposes that Fire Scenarios be made available to all components of our educational system and to appropriate public forums.

4.4.1 Public School

Based on its combined experience, the committee finds that the nation's educational system does not include adequate instruction on fire safety. Schools that provide courses in driver education, sex education, and health education give little or no instruction in fire education. Possible exceptions are the talks and demonstrations given during National Fire Prevention Week and the fire drills which most schools conduct. This deplorable situation exists despite the fact that virtually all children return to homes that contain considerable combustible material.

A series of specific fire scenarios could be developed that alert young citizens to the potential for fire in their homes. Intelligent discussion of such scenarios should bring about a greater awareness of the needs for fire safety.

4.4.2 Trade Schools

In lieu of education for entrance to College, many municipalities have developed vocational schools that train young people for careers as plumbers, electricians, mechanics, machinists, carpenters, secretaries, etc. A review of their curricula indicate little training in Fire Technology. Since new construction is utilizing increasing amounts of new synthetic combustible materials, training in Fire Technology for persons in the building trades is becoming essential. Moreover, the complexity of functional interrelationships among workers in the various aspects of construction, dictates the use of Fire Scenarios as training aids.

4.4.3 College Level

In colleges with fire-related research programs, some students are exposed to certain technical aspects of fire prevention and control. However, the majority of college students are in curricula devoid of any fire related training.

In colleges the scientist, engineer, salesman, businessman, financier and lawyer, are given their first levels of professional education so that they may work effectively in society. And no matter what occupation a student chooses, fire will be a factor training in this area. Fire scenarios can be used to illustrate the role of each profession and its interaction with others and fire.

4.4.4 Continuing Education

While the introduction of the Fire Scenario Method into regular educational curricula could have long term effects, doing so would not be enough. Fortunately, continuing education courses, emphasizing fire scenario analysis, could be designed to show interactions between private and public organizations, trades and professional organizations, in addition to imparting specific knowledge to individuals, consistent with recent congressional action to bring about greater fire safety. The continuing education system is ready made and widely accepted. It could be utilized for instruction in how to prevent and control fire.

4.4.5 NFPCA National Fire Academy

The new NFPCA National Fire Academy, which is to serve as the focus for education of the fire services, is potentially the most important educational institution in America through which the Fire Scenario Method could reach the broadest audience.

4.5 Analysis of Fire Scenarios

The following subsections are illustrative of how fire related topics might be covered in a fire scenario analysis. Guidelines for analyzing fire scenarios are presented in Chapter 5.

4.5.1 Design — Materials Selection

Development of more fire resistant materials is advancing in response to growing concern about new legislation and more stringent codes. At least for the immediate future, designers and architects are faced with the problem similar to that recently confronting officials of the new Washington Metropolitan Transportation Authority (WMTA) Subway System. They selected materials consistent with the codes and standards at the time of selection, but, when it came time to make the system operational eight years later, new information on fire had emerged that made it advisable to reevaluate their design and material choices. The dilemma, posed by higher costs and the diversion of available money, is that of possibly impaired performance of the system.

Basic to this problems is the lack of methodology for making design choices and material selections. The Fire Scenario Method could supply an essential part of such methodology.

In choosing materials, the designer should consider applicable fire scenarios in which contemplated materials may be involved. He might ask, what are common ignition sources in this application? Will such sources easily ignite the material? Will the material enhance the flame spread? Will the resulting fire readily cause the ignition of a secondary material? Will smoke production be enhanced by the presence of the material? Will flashover occur earlier than usual in fires in which these materials might be involved? Would a design modification reduce the hazard?

4.5.2 Regulations

Fire scenarios, combined with statistics and fire dynamic considerations, form a valuable resource in assessing the probable effectiveness of changing existing regulations, codes and statutes. It's too simplistic to state that, if the codes were followed, fire would not occur; or that codes need revamping because fires happen. Fire scenarios should form a part of the rationale for the development or rewriting of regulations and codes.

4.5.3 Fire Detection Systems

Until recently, humans and animals were the only reliable fire detectors available. The fire record of this country gives statistical support to the fact that sole reliance on them has not been satisfactory. For such reasons, automatic detection equipment is available, in some areas required, and additional equipment is being developed. However, the public needs to be informed about the capabilities and limitations of the new detection systems. Builders, architects, electricians, heating and air conditioning experts, insurance company personnel, city and town managers, as well as fire fighters need to understand how detection systems work to obtain the maximum benefit of such devices. The Fire Scenario Method should allow an analytical comparison of the relative merits of the systems available. In addition to a comparison of existing systems, such analyses should suggest improvements for future systems. Fire scenario analysis could also indicate when automatic fire suppression systems would be most effective.

4.5.4 Containment

Containment of a fire becomes progressively difficult as a fire develops. For example, it is easy to strike a match, use it, and blow it out. The burning match, when dropped in a forest, can ignite leaves to create a small fire which can readily be stamped out. However, as a small fire moves away from the ignition source, it often gets beyond the capability of those at the point of initiation. From there on its course and outcome will be determined by the wind, weather, and terrain until it is contained by some boundary circumstance.

Clearly, there are points in the growth of fires at which man can actively and directly stop their spread, while there are other points at which it may be better to prevent fire spread by barriers such as the walls of a building or moats in a field. Active containment, i.e. preventing a fire from growing, is accomplished through use of inerting materials (e.g., water, foams, and chemicals) at the periphery of the fire.

How effective are such methods? Through the use of fire scenarios, it should be possible to determine when a particular coarse of action can be taken and how it could be improved upon. Obviously, the lower the rate of flame spread a material demonstrates in a fire environment, the more easily it may be controlled because one has more time to gather extinguishing substances to contain it directly or to call for help. Noncombustible materials would be the ideal, but to date such materials appear unlikely to be used in competition with polymers because of the many esthetic, economic, and functional advantages of the latter. Accordingly, what rate of flame spread in a given situation can be coped with using available containment techniques? A simple example is found in the case of fiberglass boats, many of which burn so rapidly that harbormasters plead not so much for noncombustible boats, as for boats that will burn as slowly as their wooden predecessors. By the use of fire scenarios one can begin to determine flame spread rates which are acceptable with respect to human containment capabilities in various fire situations.

4.5.5 Life Support

When the ability of an individual to operate within a fire environment varies as a function of distance from the fire, the size of the fire, and the combustion products produced in the fire, considerable judgement enters into the provision of life support equipment. How much equipment would an individual use? How much time does an individual have to get into such equipment? Is it better to risk a little harm to an individual so that a total system can function? Circumstances have arisen in which the complete protection of individuals led to disaster for the system because the operators could not see each other; a factor which was realized only after life support apparatus was put into use.

Fire Scenarios analysis employing human engineering techniques can be helpful in determining what kind of equipment to use for coping with a particular type of fire. From scenarios one should be able to select the types of equipment required for generic classes of fire.

4.6 Conclusions and Recommendations

Fire, a continuing and growing problem, poses complex problems that will not be solved by a singular attack on materials. Fire hazards are not the result of any one group's activities, but rather result from the actions of several groups which, for the most part, rarely interact with each other. Furthermore, such groups make their contributions to a fire long before it actually occurs.

The Fire Scenario Method (or its equivalent) is recommended as the means to develop an integrated approach to coping with the fire problem. It is also recommended that fire scenarios be generated and made available for use in the educational system. It is suggested that analyses of these scenarios will increase public awareness of fire, and provide personnel with responsibilities for materials selection and design a methodology for decision making in their work.

CHAPTER 5

GUIDELINES FOR DEVELOPING AND ANALYZING FIRE SCENARIOS

5.1 Introduction

This chapter is concerned primarily with the consideration of important facts about a fire which ideally belong in a scenario. Although it is recognized that virtually all real fire investigations are handicapped by the absence of trained observers, especially at the early stages of the fire, and that it is frequently necessary to guess what happened from fragmentary evidence, it is useful to indicate here with information is desirable. In some cases, it is desirable to set up a simulation of a fire scenario: to determine whether or not what one thinks happened could really happen, to instrument the fire and obtain quantitative data on critical fire dynamic elements, to investigate changes in the scenario from design modifications and/or the substitution of different materials, etc.

In the following discussion of fire scenarios, the physical behavior of the fire is emphasized, but the interactions with humans are deemphasized because this study is directed at fire safety via modifying materials rather than people. Nevertheless, the action of the people may enter into a fire scenario by involvements such as: starting the fire, preventing the fire, detecting the fire, extinguishing the fire, escaping from the fire, and being killed or injured by the fire. Moreover, the psychological and physiological aspects of human involvement are beyond the scope of this report.

Judgement and extrapolation are very important in developing scenarios because only very limited data are available and technological change may occur so rapidly that the time lag between the introduction of new materials, products, or structures and the development of statistically significant accident histories may be unacceptable.

5.2 Pre-Fire Situation

In general, important events in the fire scenario occur long before the ignition source sets the fire in motion. Frequently, decisions made during the planning, design and building of a structure will profoundly affect the subsequent events in the fire chain. Therefore, it is essential that proper attention be given to the pre-fire situation since in some instances the optimum solution to the fire problem will result from action taken long before the fire begins.

Thus, the first step in the development of a fire scenario should include the gathering of data such as local building codes, governing regulations, plans and specifications, builder's and manufacturer's records and inspection records. Attention should be directed to the basis for materials selections, how and where the materials were to be used, and how the materials were installed. Specifically, did the materials meet the applicable codes; were they used properly and were they installed correctly?

5.3 Ignition Source

Appropriate information about the ignition source is required to characterize it quantitatively because, in many instances, ignition of the target fuel is marginal, i.e., embers fall on a rug but self-extinguish or a blowtorch impinges on a plywood panel but only chars it.

The primary characteristics of the ignition source are:

maximum temperature
energy release rate
time of application to target
area in contact

In some cases, details of the heat transfer mechanism from the ignition source to the target fuel must be known because it may be a critical combination of conduction, convection, and radiation. The degree of air motion or turbulence may influence spontaneous ignition of a heated vapor rising from a surface. Access to oxygen is also important. Alternatively, a target immersed in hot combustion products may not ignite because oxygen is excluded by the heat source itself.

The most important single fact to recognize about a potential ignition source is that, for solid polymers which are not readily ignitable, a "strong" ignition source will generally ignite the target while a "weak"one will not. The "strength" of the source depends on the energy flux and on the time of application to the target, or sometimes on the product of these two factors.

5.4 Ignited Material

The first material to be ignited is important in the scenario because the probability of ignition occurrence depends on the properties of the target material. If the target material is flammable liquid, its ignitability will depend on whether it is the form of a stationary pool, a foam, a mist, or a spray. For a stationary pool, the initial temperature is crucial. If its temperature were below the flash point, ignition would occur only after sufficient heating to bring a substantial portion of the liquid to the fire point. If its initial temperature were above the fire point, ignition of the fuel vapors above the pool would occur immediately and the pool would easily sustain burning.

If the target material were a gas mixture, ignition would depend primarily on the thermodynamic properties of the gas (composition, temperature, pressure). Moreover, the gas mixture would have to be heated over a region which is large comapred with the quenching distance of the mixture; i.e., a very fine wire or a very small spark would fail to ignite the mixture even though the temperature of the ignition source were far above the ignition temperature of the mixture as conventionally measured. A gas mixture in rapid motion is harder to ignite with a weak source than a stationary gas mixture.

In most fire scenarios, the target material is solid. Ignition of solids results from ignition of the pyrolysis products which are evolved during thermal decomposition of the solid, and occurs in the gas phase above the solid surface. The ignitability of a solid depends not only on its chemical composition, but also on an energy balance at the surface (including radiation), on its thickness and thermal properties, on its configuration and on gas phase conditions.

When considering the chemical composition of a target material, the following factors are especially relevant: 1) basic material may contain small percentages of additives (fire retardants) or impurities, which may have major effects on ignitability; 2) if the material is hygroscopic, like cotton, the initial moisture content will vary over a wide range depending on prefire humidity, with important effect on ignitability; 3) if the material contains several major constituents, such as flexible polyvinyl chloride containing a large proportion of plasticizer, its ignitability will depend on the more volatile constituent, in this case the plasticizer; 4) the target may be composite in nature, consisting of an outer skin material and an underlying material, either of which may contribute to ignitability.

The importance of energy balance at the surface of a target is shown by attempting to ignite a single piece of wood such as a two-by-four. Self-sustained burning will not result unless the ignition source is applied for a very long time, so that the average temperature of the wood reaches about 320 °C. Yet, a match placed between two vertical two-by-fours close together can start self-sustained burning. The single piece of thick wood cannot continue to burn because of the high rate of radiant energy loss from the charred hot surface to the cold surroundings. This effect is less important for materials which burn at lower surface temperatures, such as non-charring thermoplastics. Yet, since radiant input from the ignition source can be important, the reflectivity of the target material is a significant factor in such cases.

The thickness and thermal properties of a material are vital in determining the time required to achieve ignition, when a given heat flux is applied to the surface, and they become crucial in the scenario if the heat flux is of relatively short duration. As previously indicated, a distinction must be made between "thermally thick" and "thermally thin" materials because the time to ignition for "thermally thick" material is dependent on the thickness and controlled by the "thermal inertia," which is the product of the thermal conductivity and the heat capacity per unit volume. For a "thermally thin" material, the time to ignition is proportional to the product of thickness and heat capacity per unit volume (fabrics generally fall within this category). Whether the material behaves in a "thermally thick" or "thermally thin" manner depends not only on the thickness, but also on the heating rate, the heating time, and the "thermal diffusivity," which is the ratio of thermal conductivity to heat capacity per unit volume.

In the case of a thin flammable material (carpet, paneling, etc.) in thermal conduct with an underlying material, the thermal properties of the underlying material can influence the ignitability by the degree to which the underlying material acts as a heat sink.

The configuration of the target material can also be of great importance. Whereas the foregoing discussion implied a one-dimensional geometry, ignition actually tends to occur more readily in a crevice or fold, or at an edge or corner, rather than in the middle of a flat surface.

5.5 Flaming or Smoldering Combustion

Some combustible materials may burn either in a smoldering mode, like a cigarette, or in a flaming mode. A material also may smolder for a certain length of time and then spontaneously burst into flame.

In general, only solids with very low thermal conductivities such as porous solids, or thin solids not in contact with a heat sink, such as suspended cotton thread of a free-standing piece of paper, can smolder. Smoldering seemingly occurs only in char forming materials (a sofa cushion made of polyurethane or foam rubber under a synthetic fabric cover can burn in the smoldering mode). Smoldering is characterized by much lower spread rates than flaming combustion.

Smoldering is important in that: a) the smoke or gases produced may permit detection of the fire at an early stage; b) the pyrolysis products may be toxic; and c) a transition to flaming, after a long period of smoldering, may produce a very rapidly growing flaming fire because of the preheating of fuel and accumulation of combustible gases during the smoldering period.

It is known that the character of smoke produced in flaming combustion is different from smoke produced with smoldering combustion of the same fuel, as shown by the differing response characteristics of smoke detectors. Consequently, this difference must be taken into account when selecting smoke detectors.

Smoldering may continue for a very long time. For example, barrel of sawdust might smolder for more than 24 hours. Therefore, scenario analysis should consider the possibility of a long time lag between ignition and active flaming as a function of the target materials.

The burning of charcoal is usually referred to as glowing combustion rather than smoldering. Its importance to the fire scenario is that cellulosic materials, after the flaming combustion is finished, continue to glow for a substantial time as the residual substance is consumed. During this time, the possibility of a resurgence of the fire exists.

Furthermore, when a gaseous extinguishing agent such as carbon dioxide or a halocarbon vapor is applied to a fire, flaming combustion may stop, but smoldering combustion might continue (deep-seated fire). After a while, the extinguishing vapor may dissipate and the flame may rekindle. Thus, a "one-shot" gaseous extinguishing system may not assure protection unless the fire is held in check long enough for effective manual extinction.

5.6 Fire Spread

5.6.1 General

Unless a person were wearing or sleeping on the originally ignited item, the fire is not apt to do much damage until it has grown by spreading some distance from the point of ignition. The rate of spread is very important in the scenario, because it determines the time after ignition when the fire reaches a dangerous size. The "dangerous size" may relate either to the rate of generation of toxic and smoky products or to the difficulty of extinguishment. The ability of people to detect, fight, or escape from a fire depends on the time for it to reach a dangerous size, and hence on its spread rate.

Fire may spread either between contiguous fuel elements or by jumping across a gap from the initially ignited material to a nearby combustible item. These two cases are discussed separately as follows:

5.6.2 Fire Spread Over the Initially Ignited Material

The rate of flame spread over a solid surface in the horizontal or downward direction is often quite slow, sometimes as little as one inch per minute. However, if the material is "thermally thin," or has been preheated by radiation or convection from hot combustion products, the flame can spread quite rapidly. If the fuel is arranged to permit upward flame propagation, flame spread will occur very rapidly and at a progressively accelerating rate. If the fuel is arrayed as a lining of a corridor or duct, with the air supply coming from the left and the combustion products existing on the right, the fire will spread rapidly from left to right until it penetrates sufficiently into the duct and most of the available oxygen is consumed. It will then stop spreading until the originally burning fuel is also consumed. Thereafter, the fire will move downstream. Thus, in such situations, the effects of ventilation are controlling.

Indeed, for any fire burning in a compartment with limited air supply, the rate of spread will decrease as the air becomes vitiated by combustion products. However, spontaneous breaking of windows, or deliberate actions of firefighters to improve visibility by ventilating the fire, will have an accelerating effect on spread rate.

Fire spread over a liquid is relatively slow when the liquid is well below its flash point, but possibly a hundred times as rapid if the liquid is above its flash point. For liquids below their flash point, motion with the liquid, induced by the fire, is important in determining the spread rate.

Flame spread through a premixed gas may vary from a foot per second to hundreds of feet per second, depending on conditions for generating turbulence by the expansion associated with combustion. For several thousand feet per second occur.

5.6.3 Fire Spread to Secondary Material

When originally burning material is separated by a gap from the nearest secondary combustible, and the flame does not impinge directly on this secondary material, the fire will die out after the original material is consumed, unless by some mode the fire can spread across the gap as discussed below:

A fire may radiant directly on the target, or convectively heat the ceiling and upper walls, which then radiate onto the target. Alternatively, hot smoky gases accumulating under the ceiling may radiate onto the target; or, hot combustible gases accumulating under the ceiling may ignite and radiate onto the target.

In any event, radiation preheats the secondary material until it pyrolyzes, emitting flammable vapors. At this point two possibilities exist. Either the secondary surface may ignite, or a sufficient concentration of flammable vapor mixture accumulates to permit the original flame to spread through it to the secondary material.

Other modes exist for fire spread across a gap. For example, if the original burning object were made of a thermoplastic polymer, it may melt to form burning droplets that fall and ignite secondary fuels they may encounter. Additionally, if an original burning object is mechanically weakened and falls over, its collapse can provide a means of fire spread. Similarly, mechanical action associated with an ineffective attempt to extinguish a fire may lead to its spread.

In the case of overheated grease in a kitchen pan fire, the burning grease may spatter droplets that spread the fire. Similarly, an ember from a fireplace log or other burning object may be propelled several feet by pyrolysis gases and cause secondary ignition. Forest fires often spread by wind-borne firebrands. The winds may be generated by the convection plume of the fire itself. Firebrands may be generated by the convection plume of the fire itself. Firebrands often contribute to the spread of conflagrations in cities, or to any outdoor fire in a high wind, as well as to fires on moving vehicles.

5.7 Evolution of Smoke and Toxic Gases

5.7.1 General

Smoke and toxic gases are important to the fire scenario because they not only may provide means for the early detection of fire, but also may interfere with visibility, escape or firefighting. They have psychological and physiological effects on humans, including confused thinking, incapacitation, and death. In most instances, deaths in a building fire are a result of the toxic combustion products present rather than as a result of the heat and flames. Furthermore, smoke may contribute to fire spread by virtue of its radiation emission or absorption; and, substantial property change may be caused by smoke or corrosive combustion products.

5.7.2 Automatic Detection

With regard to automatic detection, the first consideration is the rate of smoke movement from the fire source to the detector. Under a no-fire condition, air movement in a room or compartment is determined by forced convection for heating, air-conditioning, and odor-removal purposes; by external winds blowing through open doors and windows, or by free-convection attributed to heat sources. For very small fires, the bouyancy effects of the fire heat are negligible and smoke will follow existing air circulation patterns. When a fire becomes larger than some critical size, the hot fire plume rises to the ceiling, and then flows under the ceiling to create an entirely new circulation path in the room. If, before the fire, the upper portion of the room were warmer than the lower portion, as would normally be the case for either a heated house in winter or a house on a hot summer day, temperature-induced stratification will exist and smoke may rise half-way to the ceiling and then spread laterally. Thus, for early detection of fires, the foregoing factors are crucial in determining detector response.

A further consideration is the response characteristics of the automatic detector to the smoke. This response characteristics of the automatic detector to the smoke. This response varies with the time-dependent concentration and particle size of the smoke at the detector, the velocity of smoke past the detector as well as detector orientation to the flow, the smoke entry characteristics of the detector chamber, the operating principle of the detector (optical transmission, optical scattering, ionization), in addition to the sensitivity setting of the detector circuit, battery, voltage, etc. It is especially important to note that different combustibles, or the same combustible flaming or smoldering under different ventilation conditions, produce smoke of different particle size and hence, detection characteristics. It is also known that smoke may "age" after it is formed, i.e., agglomeration of smaller particles into larger ones will occur with consequent effects on the ease of detection.

5.7.3 Visibility

Optical scattering properties of the smoke depend strongly on particle size as well as concentration, so the vision-obscuring aspects which interfere with escape or fire-fighting are strongly dependent on the type of combustible and mode of combustion. For example, incomplete burning of polystyrene or rubber produces large soot particles capable of obscuring vision even at low concentrations.

The lachrymatory effects of gases, such as aldehydes or acids associated with the smoke particles, have been shown to be important in interfering with a vision.

Visibility at floor level will generally be much better than at higher levels in a room, so the possibility of crawling to safety is important. The heights at which exit signs should be located is relevant here. If sprinklers operate, both the cooling and entrainment effects tend to bring the smoke closer to the floor. Moreover, fog, which may result from the employment of sprinklers, will interfere with vision.

5.7.4 Toxic Effects

Carbon monoxide is the chief toxicant, according to our present knowledge, but other substances such as acrolein, HCN, HC1, HF and CO_2, which may be present in the smoke can be very hazardous in certain cases and might exhibit synergistic effects.

When escape is not possible, the critical survival concentration of toxicant depends on the time of exposure, which, in turn depends on the history of the fire. Also, combined effects of toxicants with heat, excitement and loss of vision are believed to be as important in determining human survival, as is the health of a subject and that individual's previous intake of alcohol or drugs. Confused mental processes, induced by toxicants, may be particularly detrimental to survival where the subject must make rapid and correct decisions on scape tactics.

5.8 Extinguishment

At some point in the development of most scenarios, either manual or automatic extinguishment activity starts. Such activities may involve smothering the fire by applying water or other agent. Techniques of extinguishment are outside the scope of this study, but the effectiveness of extinguishment depends on the burning characteristics of the polymeric combustible. If the fire has become too large, or is growing too rapidly at the time extinguishment is attempted, the fire will not be controlled.

Accordingly, the rate of fire spread and the maximum rate of burning fire-fighting, a critical factor is how closely can the firefighter approach the fire, and will smoke prevent him from determining where the fire is? If he has a hand-held extinguisher of given capacity, will it be enough to do the job? When automatic sprinklers are present, there is usually no problem unless the fire is shielded from the sprinklers (i.e., in an unsprinklered closet) or unless it is a high-challenge fire (i.e., polystyrene foam products stored 15 feet high, in which case the key variable is sprinkler design density).

5.9 Flashover

Flashover, a critical transition phase of a fire in a compartment, generally occurs in ventilated compartments; otherwise the fire will tend to smother itself before the flashover stage is reached. Prior to flashover, the rate of the local fire burning in the compartment is determined by the extent of flame spread to that time. After flashover, all flammable contents in the compartment are burning or rapidly pyrolyzing, flames are projecting out the door or window, and the burning rate within the compartment is determined by the rate of ventilation and/or the total exposed fuel area. Flashover often occurs suddenly, within a few seconds, and is characterized by very rapid fire spread throughout the compartment, with flames violently rushing out the door or window.

Whether flashover can occur in a compartment depends on the compartment's size and shape, ventilation available, intensity of the initial fire, as well as the quantity and disposition of secondary fuels. If flashover can occur, the time required for its occurrence will depend not only on the foregoing variables, but also the thermal inertia of the room, especially the ceiling. A fire in a typically furnished room will require five to twenty minutes after flaming ignition to reach flashover.

During the pre-flashover period, the upper portion of the room is filled with hot, smoky, oxygen-deficient combustible gases. The lower portion contains relatively cool, clean air coming from the door or window. At some intermediate height, perhaps two feet under the ceiling, there may be both sufficient oxygen and heat to readily ignite target fuels at that level. Drapes or curtains are examples of materials in this region. Occasionally the hot combustible gases ignite as they reach an adequate source of oxygen, resulting in rapid and violent combustion of the accumulated hot pyrolysis gases.

Radiation is probably of major importance in flashover. Thus, the infrared emission absorption, and reflection characteristics of objects and smoke in the compartment are highly relevant.

The larger the volume of a compartment, the less likely it is that a fire of given size will cause flashover. Data on simulated room fires indicates that, for a 12′ X 12′ X 8′ room, a fire consuming 2 pounds of fuel per minute could produce flashover in about 20 minutes, while if the combustion rate were twice as high, flashover would occur in only 1.5 minutes. Thus, one would suspect that even a very large, sparsely furnished room could flash over, if a high rate of initial burning were achieved, because the time to flashover is extremely sensitive to rate of heat release. The BOAC passenger terminal fire at J.F. Kennedy Airport, New York, in 1970, illustrates a fire in a large space with low fuel loading (but presumably a very high heat release rate because of the plastic foam padding on lounge chairs), which apparently developed in a short time to flashover condition.

5.10 Spread to Adjacent Compartments and Catastrophic Failure

Fire resistant compartmented buildings are designed with the expectation that a fire in any one compartment will be confined by the building structure itself so that either the fire is extinguished or the fuel is exhausted before the fire breaks through to adjacent compartments. Interior partitions, fire doors, etc., are subject to building codes specifying that the partition must maintain its integrity for an appropriate time (such as: one hour, two hours, etc.) depending on circumstances.

Thus, the scenario should include information on the fire endurance rating of the relevant structural elements.

Assuming that the building itself is fire resistant, the fuel loading, expressed in pounds per square foot, influences the duration of the fire once it has grown large enough to become ventilation-controlled. As a rough rule of thumb, there is an endurance requirement of about 0.1 hour for each pound per square foot of fire load (assuming certain typical ventilation rates). The fire load may range from a few pounds per square foot, in lightly furnished occupancies, to an order-of-magnitude higher fire load in storage occupancies. The endurance requirements are based on the assumption that the fuel is primarily cellulosic; however, if it is primarily a polyolefin or rubber, the stoichiometric air requirement per pound of fuel will be up three times as large and the heat release per pound of fuel will also be much higher. Moreover, a ventilation-limited polyolefin fire may burn differently than a cellulosic fire.

Of course, if the fire compartment has openings to other sections of the structure, such as open doorways, ventilating ducts, improperly fire-stopped or inadequately sealed openings in walls, etc., these would constitute critical elements in the scenario. Even if they were confined to the compartment of its origin. The spread of smoke and toxic gases throughout the structure could have catastrophic effects. If the fire were capable of heating structural elements of the building to failure (steel above 1000 °F), the structure might collapse. Thus, the thickness and integrity of insulation on structural elements also become important to the fire scenario. Of course, local structural collapse within a sprinklered building, may cause breakage of the sprinkler piping and consequent escalation of the fire damage.

In multi-level structures, it is especially important to prevent any means by which fire may spread progressively upward from each level to the next higher. Key elements in this type of scenario are: ventilation passages, utility passages, fire endurance of ceiling, stairwells or elevator shafts, and flame projecting from a window so as to cause ignition through the window of the floor above.

5.11 Spread to Other Structures

Where a structure becomes completely involved with fire, a finite probability exists that adjacent structures will ignite. Ultimately, a conflagration involving a large area may result. Such fire propagation could occur either by radiation or firebrands.

Potentially critical factors in the fire scenario are: magnitude and direction of the wind, separation distance between structures; flammability of roofing material such as wood shingles, ignitability by radiation of curtains inside windows facing fire, combustible trash in alleys between buildings, and propulsion of burning debris after building collapse or explosion.

Since the spread of fire to other structures usually occurs at a sufficiently late stage of a fire, firefighters will probably be present. Their tactics in wetting down adjacent buildings are extremely valuable in preventing spread to such structures. Conversely, if the fire were simultaneously burning in many areas, as could be the case for a brush fire or fire caused by civil disorders or military incendiary attack, firefighting will probably be inadequate, and the degree of spread will depend on the intrinsic "fire-hardness" of the structures involved.

Similarly, if a fire were started by the effects of an earthquake or strong explosion, water mains would probably be broken and firefighting would become ineffectual. In that case, spread would again be limited by the intrinsic "hardness" of structure involved, which is material-dependent.

5.12 Essential Fire Scenario Elements

A scenario should cover as many as possible of the following points.

(a) The pre-fire situation.

(b) The source of the ignition energy should be identified and described in quantitative terms.

(c) The first material ignited should be identified and characterized as to chemical and physical properties.

(d) Other fuel materials that play a significant role in the growth of the fire should be identified and described.

(e) The path and mechanisms of fire growth should be determined. Particular attention should be given to fuel element location and orientation, ventilation, compartmentation, and other factors that affect fire spread.

(f) The possible role of smoke and toxic gases in detection, fire spread and casualty production should be determined.

(g) The possibility of smoldering combustion as a factor in the fire incident, e.g. as a cause of re-ignition, should be considered.

(h) The means of detection, the time of detection, and the state of the fire at the time of detection should be described.

(i) Defensive actions should be noted and their effects on the fire, on the occupants, and on other factors should be described.

(j) Interactions between the occupants of the building or vehicle and the fire should be detailed.

(k) The time sequence of events, from the firstoccurrenceof the ignition energy flux to the final resolution of the fire incident, should be established.

The scenario should permit generalization from the particular incident described. It should provide a basis for exploration of alternative paths of fire safety performance of changes in materials, design, and operating procedures.

5.13 Analysis of Fire Scenarios

5.13.1 General

While the analysis of any fire scenario might be accomplished in several ways, an effective procedure is to ask questions about each essential fire scenario element. The answers to such questions should suggest means for fire prevention and control, as well as provide bases for making materials selection, establishing design criteria, validating test methods, promulgating codes and standards, and formulating research and development objectives. Typical of the questions that might be asked are as follows:

5.13.2 Pre-Fire Situation

1. Were existing codes adhered to? If so, did they yield adequate performance? If not, why not, and would they have been effective had they been enforced?
2. Were materials installed properly? If so, did they contribute to fire growth or did the help contain it?

5.13.3 Ignition Source

1. In as much detail as possible, what was the ignition source?
2. For how long was it in contact with the ignited material prior to flaming ignition? If this is not known, could it be determined by a separate experiment?
3. Could the ignition source be eliminated? How? By education? By design?

5.13.4 Ignited Material

1. What was the originally ignited material? If composite, what were the various layers?
2. What was the application of the material (e.g. drape, rug, cushion)?
3. How was it located relative to ceiling and nearest wall?
4. What were the ventilation conditions in the room?
5. What was the relative humidity during the 48-hour period before the fire?
6. Did melting and dripping of the ignited material occur? Did this factor significantly affect the fire spread?
7. Did the ignited material collapse, fall over, etc.? If so, what effect did it have on the fire?

8. Once the ignited object is fully involved, is it possible to quantitatively estimate its heat release rate?

9. Are there other materials that could have been employed in this application which would not have ignited under the same exposure conditions? If so, why weren't they employed?

10. Do flammability tests on maerials intended for this application adequately measure ignition resistance to this level of ignition source? Should they?

5.13.5 Flaming or Smoldering Combustion

1. Is it known if smoldering preceded flaming? For how long?
2. If unknown, was the ignited material capable of smoldering?
3. Can the volume of gases produced by smoldering be estimated?
4. Can the time-dependent concentration of smoke and toxic gases arriving at a strategic location some distance from the fire be estimated?

5.13.6 Fire Spread

1. How long did it take for the first ignited object to become fully involved?
2. If flame spread to a second object, what was the mechanism of energy transfer?
3. How was flame spread influenced by sudden events, such as breaking of windows, opening of doors, melting and dropping of curtains, spattering of burning droplets, etc.?
4. Did one or two materials significantly control the fire spread rate? Could the substitution of different materials, or the incorporation of design modifications, alter the rate of fire spread and growth?

5.13.7 Smoke and Toxic Gases

Automatic Detection

1. If a smoke detector were present, how was it located relative to the fire? Did it respond as would be expected?
2. Supposing no smoke detector were present, how much sooner would the fire have been detected if a smoke detector had been present in a logical location? Would detection have been soon enough to make a crucial difference?

Visibility

1. Was visibility obscured in an escape route: When did this obscuration occur relative to detection time? Which materials seemed to contribute significantly to visibility obscuration?

Toxic Effects

1. Were victims affected by toxic substances?
2. What toxic substance caused death (autopsy)?

3. Did toxic substances interfere with victims' escapes by promoting confused thinking or decision making?

4. Could these toxic substances be attributed to any one material?

5. Were victims affected by pre-existing conditions such as limited mobility, circulatory disease, recent alcohol or drug intake, etc.?

5.13.8 Extinguishment

1. How large was the fire when first detected? What were visibility conditions at the time?

2. How much time elapsed between detection and attempted extinguishment?

3. How large was the fire when extinguishment was attempted?

4. What was the extinguishment technique and how successful was it?

5. If automatic sprinklers had been present, how much sooner would they have been expected to control the fire, and how much less might the loss have been?

5.13.9 Flashover

1. Did flashover occur? How long after ignition? How long after detection?

2. Can crucial elements in the fire growth and spread process be identified in relation to flashover?

5.13.10 Postflashover

1. Did fire spread beyond the initial compartment? How? Was door opoen?

2. How was the ventilation system involved in fire spread?

3. Did walls, fire doors, etc., fail? If so, after how long?

4. Did Fire spread to floor above? By what mechanism?

5. Did structural collapse occur? Was it due to code violations?

5.13.11 Spread to Other Structures

1. Did other structures ignite? How far away were they?

2. Was radiation responsible? Were firebrands involved?

3. What were wind conditions?

4. Did firefighters attempt to protect exposed structures? How soon before spread occurred?

5.13.12 Summary

The availability of accurate, detailed, and complete fire scenarios allows opportunity for an indepth fire hazard analysis, as illustrated by the preceding questions. As such questions are raised, and some are answered, means for fire prevention and control emerge. These means may involve more education, better material selection, improved designs, installation of detection equipment, more stringent codes, etc. Whatever solutions emerge, they should be based on a comprehensive overall systems analysis of the fire problem enhanced by fire scenario analyses.

The scenario concept is a common tool in long-range planning[1]. However, only recently has the scenario concept been applied to fire-safety program planning. In 1976, the national Bureau of Standards' Center for Fire Research employed 5,040 different fire scenarios in developing a research plan to reduce the Nation's fire losses[2]. The National fire Protection Association in its publication *Fire Journal* documents the chain of events in many actual fires which they investigate. These real-life fire scenarios have been employed as the basis for in-depth studies of fires in specific residences, e.g., one- and two-family dwellings[3], mobile homes[4], and nursing homes[5]. Other than these instances, there seems to have been little use of the fire scenario approach to improve fire safety.

5.14 Conclusions

1. Although all fires differ, the development and analyses of fire scenarios lead to the identification of their common elements. Such analyses also allow identification of the critical stages in fire development, suggest opportunities for fire prevention, and direct attention towards various methods for control.
2. Scenarios should be prepared to permit generalization from the particular incident described.
3. A good scenario provides a basis for the exploration of alternative paths of fire initiation and growth as well as for analysis of the effect on fire safety perofmrnace of changes in materials, design, and operating procedure.
4. The application of fire scenario analyses appears to be a most productive methodology to identify effective means to improve fire safety vin our increasingly complex environment.
5. At present, insufficient use is being made of the fire scenario technique to advance fire safety.

5.15 Recommendations

1. A broad spectrum of generalized fire scenarios should be developed that are based on real or credible incidents. Specific fire dynamic elements in such scenarios, e.g., rate of fire spread, rate of heat release, etc., should be further quantified when necessary by information obtained from large scale experiments. The ultimate goal of this recommendation is to develop the capability of employing analytical methods to predict fire hazards and thus replace expensive large scale experiments.
2. Scenarios should be used for the analysis of fire hazard and for the development of methods to provide increased safety. In particular, scenarios should provide the basis for materials selection, design criteria, validation of test methods, promulgation of codes and standards, and formulation of research and development objectives.
3. The design and procurement of any system, subsystem or structure, employing polymeric materials, where fire safety is a design consideration, should be based on the development and testing of the design against appropriate fire scenarios. In essence, the recommendation, if adopted, would require the development of fire impact statements.

4. Fire safety personnel should be trained in the development and use of fire scenarios, so as to enable them to identify critical fire hazard elements and determine appropriate protective measures.

References

1. Zentner, R., ''Scenarios in Forecasting'' Chemical and Engineering News, Vol. 15, No. 40 (October 6, 1975), p. 22 and references cited therein.
2. ''Reducing the Nation's Fire Losses, the Research Plan'', Center for Fire Research, National Bureau of Standards, January, 1976.
3. A Study of One-and Two-Family Dwelling Fires, NFPA No. FR75-1, National Fire Protection Association, Boston, Mass. (1975).
4. A Study of Mobile Home Fires, NFPA No. FR75-2, National Fire Protection Association, Boston, Mass. (1975).
5. Nursing Home Fires and Their Cures, NFPA No. SPP-17, National Fire Protection Association, Boston, Mass. (1972).

CHAPTER 6

SPECTRUM OF FIRE SCENARIOS

6.1 Introduction

This chapter presents a variety of brief scenarios, based on real fire incidents in which polymeric materials play a significant role, to illustrate the diversity of fires, and show the commonality that permits a scientific approach to fire safety. In these scenarios, attention is directed to the following areas: prefire environment, ignition source, material first ignited, other significant materials involved, fire dynamics, method of detection, extinguishment, and extent of loss.

6.2 Aircraft Fires

6.2.1

A transatlantic jet airplane was approximately a half hour from its destination when a passenger went to the rear of the plane and opened the lavatory door. He was greeted with a wave of white smoke. He quickly closed the door and called the attendant. The attendant reported the situation to the captain who sent the flight engineer back to investigate, at the same time requesting permission for emergency descent. The flight engineer found the rear of the cabin filling with black smoke. He instructed the cabin steward to attack the fire with portable CO_2 extinguishers while he increased the air flow to the cabin to keep the smoke in the rear of the airplane. The fire extinguishers were ineffective. Within three minutes of the discovery of the fire, the smoke had reached the cockpit and was interfering with visibility. The cockpit windows were opened, making visibility possible, and the pilot prepared for an emergency landing. The aircraft came to rest in a field with the fuselage practically intact, but a fire broke out in the cabin and spread through the cabin in seconds. There were only five survivors. It is believed that most of the passengers were overcome by smoke as well as toxic gases and would have been unable to evacuate the plane without assistance, even if the fire had not intervened to prevent rescue operations. The cause of the fire could not be determined, but a discarded cigarette in the lavatory trash container is suspected.

6.2.2

A jumbo jet, having made a cross country flight, was given a brief post flight inspection and secured for departure on the following day. Approximately four hours later an airline employee smelled smoke next to the aircraft passenger ramp. Upon boarding the plane he noticed smoke emanating from the rear of the cabin area. He immediately reported the fire to the airport fire department. They responded quickly and found fire burning through the fuselage. An attack was made, and it was realized that additional aid was needed. This was requested. Approximately 20 minutes had elapsed from the initial reporting of the incident to burn through of the cabin area fuselage. The entire cabin area was destroyed. Post-fire inspection revealed that the ignition was caused by a short attributed to routine maintenance in an electric outlet in the lavatory which ignited paper toweling in a waste container. The fire then traveled up, between fuselage frames to the overhead ceiling area, and then forward. Burned wiring bundles indicated the intensity of the fire above the ceiling panels. Besides paper toweling and tissues, combustibles contributing to the fire included acrylonitrile butadiene styrene (ABS) or vinyl-type thermoplastics, wiring insulation, wood frames for the vertical and ceiling panels, and the neoprene/nylon vapor barrier covering the insulation blankets. The fire was finally extinguished by the airport and municipal fire departments.

6.2.3

Shortly after a jumbo jet aircraft started down the runway for takeoff, a severe power reduction was noticed. It seems that small birds, endemic to the surrounding airport, were ingested by the jet's huge air intakes. The pilot aborted the take-off and the plane came to rest just off the end of the runway with minor damage, but the hot engine ignited jet fuel that spilled after the crash. All occupants of the aircraft successfully exited down the plane's escape chutes; however, within 90 seconds after the crash, the entire right side of the plane was engulfed in flame. Despite rapid response of the crash rescue services, the side shell of the main cabin was breached and the interior ignited. This interior ignition rapidly burned several large holes in the fuselage and served as a continuing source of ignition to flammable vapors escaping from the plane's wing tanks. Immediately upon breaching by the external fire, the interior of the fuselage became filled with smoke, which could have seriously hampered evacuation.

6.2.4

At approximately one hour from destination of a transcontinental flight, a stewardess reported to the flight deck that there was a fire in the galley. Electrical power to the galley was cut off from the cockpit. Smoke and flames were observed in the vicinity of the coffee maker. The crew used a CO_2 extinguisher and two dry powder extinguishers to put out the fire. The flight engineer removed the coffee maker from the galley. The galley interior was charred, but there was no other physical damage. The flight continued to its destination without further food service. Subsequent examination of the coffee maker revealed that a short circuit had ignited the plastic housing.

6.3 Building Fires

6.3.1 Residential Buildings

6.3.1.1

Two couples had been playing cards in the downstairs reception room of a two story frame house. About midnight, the visiting couple left and the residents went to bed in an upstairs bedroom. At approximately 2:00 A.M. a neighbor saw flames at the reception room window and called the fire department. He attempted to rouse the occupants of the burning house, but received no response. The fire department responded within five minutes and found the recreation room fully involved. The fire, which had not spread beyond the recreation room, was quickly extinguished, but the occupants were found in their bedroom, both dead. The man was lying back on the bed with his feet on the floor; the woman was at the bedroom door. Smoke stains on the wall indicated that smoke was carried into the upper part of the room by the air conditioning system and accumulated down to a level of four feet from the floor. Examination of the recreation room revealed that an upholstered sofa was almost completely destroyed, draperies at the windows had burning igniting an adjacent plywood panelled wall, and a mineral acoustic tile ceiling was badly discolored, but had not contributed to the fire. The wall-to-wall carpeting was burned in the neighborhood of the sofa and draperies, but had not contributed to the spread of the fire. It is probable that a discaded cigarette, left behind at the end of the card game, ignited the sofa, which smoldered for some time, producing large quantities of smoke and toxic gas before bursting into flame. The limited extent of fire damage indicated that flaming combustion had been in progress for only a short time before the fire was extinguished. Death was attributed to carbon monoxide poisoning.

6.3.1.2

When cleaning up after a later evening party, ash trays were emptied and other trash was collected in a plastic trash bag. The bag was placed in a plastic trash can in the attached garage of the single family home. The occupants retired shortly thereafter. About an hour later, they were awakened by the alarm of a smoke detector located in the central hallway of the house. Smoke odor was evident. Upon investigation, smoke was seen entering the kitchen around the door leading to the garage. Upon carefully opening the door, flames were observed in the garage. The door was closed and the fire department called. When it arrived, the garage was fully involved and smoke and flames were coming out of a window on the side away from the house. The fire was quickly extinguished. A fire resistant wall had prevented spread into the house. Apart from minor smoke damage, damage was limited to the garage. A burning cigarette had ignited trash in the trash can. The can melted, collapsed, the lid came off, and the plastic material of the can caught fire. The fire spread to adjacent trash cans and then to other combustibles in the garage.

6.3.1.3

A fire started at about 1:30 A.M. in a sixth floor apartment of a twelve story apartment building for the elderly. The cause was unknown, but the occupant of the apartment was known to be a chain-smoker. The occupant got out of the apartment but collapsed in the corridor outside, leaving the apartment door open. The body was found near the open door after the fire was extinguished. The building automatic smoke alarm alerted the attendant on duty at the front desk. He took the elevator to the sixth floor to investigate. His body was found later in the sixth floor elevator lobby just in front of the elevator. At 1:45 the fire was reported to the fire department by telephone from a neighboring apartment building. When the firefighters arrived, they found that the fire had spread from the room of origin down the corridor over the carpet. A one and one half inch gap for ventilation under the doors of other apartments not only allowed limited fire spread over the carpet into those apartments, but also permitted smoke and gas penetration. In addition to the occupant of the room of origin and the building attendant, five other persons died of carbon monoxide poisoning in their apartments on the sixth floor. Fire resistant construction prevented structural damage and spread of fire to other floors.

6.3.1.4

As a housewife was preparing breakfast and reached for the teakettle at the back of her stove, the loose sleeve of her housecoat brushed the front burner. The garment caught fire. Her screams alerted her husband who rushed into the kitchen, threw her onto the floor, and smothered the flames. The woman received severe burns over twenty percent of her body. She was hospitalized for two months and is permanently disfigured. Her husband received minor burns. Damage to the house was slight. The incident was not reported to the fire department or the fire insurance company.

6.3.2 Non-Residential Buildings

6.3.2.1

An electric coffee maker was left "on" in a ninth floor file room of a high rise office building. Around 9:30 P.M., the building fire alarm system indicated smoke on the ninth floor. The fire department was called and a building guard went to investigate. On entering the office suite, he found a great deal of smoke and flames coming out of the file room door. He closed the outer door and went to meet the firemen. The fire in the file room was quickly extinguished with hose lines. It was found, however, that the fire had penetrated a louvered door into a telephone closet where burning cable insulation was spreading the fire to other floors through the open cable shaft. The fire was confined to the cable shaft, but extinguished with difficulty. Several firemen suffered from "smoke inhalation." The ventilation system, which had been switched from a recycle mode top 100% fresh air when the alarm was given, helped to prevent the spread of smoke through the building.

6.3.2.2

Two construction workers were using a torch to cut holes in the interior wall of an unsprinklered refrigerated warehouse to install a new conveyor. The wall was of steel construction with three inches of sprayed on plastic foam insulated on the inside. Shortly after the work was started, the workmen observed fire in the interior of the warehouse around the hole they had made. They entered the warehouse through a door and found the insulation burning. One went to call for help while the other attempted to fight the fire with a portable fire extinguisher. When the first one returned to the scene, black smoke was pouring from the door. He was unable to enter the warehouse or locate his companion. After the fire department arrived, firemen wearing self-contained breathing apparatus quickly extinguished the fire. The body of the workman who had tried to fight the fire was found near the door. He was unburned. A portion of the insulation had been burned from the wall, but there was no structural damage to the building. Restricted ventilation had limited the growth of the fire.

6.3.2.3

A plant that manufactured expanded polystyrene egg cartons from polystyrene pellets, used a process that involved extrusion with a blowing agent, followed by thermal forming in presses. The egg cartons were packed 200 per polyethylene bag, with 18 bags stacked on a 4′ X 4′ piece of cardboard, making a 6′ high stack. The storage array was 3 stacks high, or 18′ to the top. The storage area was sprinklered at the ceiling level. Ignition was caused by an overhead lighting fixture in contact with the stored bags. The fire burned with such intensity that sprinklers, at 0.4 gallons per minute per square foot, were unable to control the fire. The fire department arrived in ten minutes, but the dense smoke hampered its operations. Three firemen were overcome by fumes from the burning plastic. After three hours of burning under conditions of restricted ventilation, a large portion of the roof over the fire collapsed. Rapid growth of the fire followed, driving the firemen from the building. Molten polystyrene covered a 10,000 square foot area to a depth of 5 inches on the floor. Damage to the building, machinery, and stock totaled $1.9 million. Production was interrupted for 8 months, at a loss of $1.5 million.

6.3.2.4

A fire of incendiary origin was started in a pile of trash that was separated by a concrete retaining wall from the seating section of an unoccupied sports stadium. No fire protection was available in this open concrete structure. Although the fire department was called immediately upon discovery of the trash fire, by the time it arrived, the fire had leaped over the wall and ignited the seats which were composed of molded plastic back and seats in metal frames. The fire spread rapidly, aided by air flow up the sloping sides of the stadium. An entire section of seats was destroyed, but the break between sections helped the fire department contain the fire. Prompt intervention by the fire department prevented all the seats in the stadium from being destroyed.

6.3.3 Custodial Buildings

6.3.3.1

A faulty Christmas tree light ignited a plastic tree in a hospital lobby. The fire spread to plastic decorative plants and then to plastic covered foam upholstered furniture. An attendant saw the fire and sounded the alarm. Fire doors prevented the heavy black smoke from traveling to other parts of the building. The fire department responded quickly and extinguished the fire. Fire resistant interior finishes on walls and ceiling, as well as a carpet that passed the current mandatory flammability standard ("pill" test), prevented spread of the fire from the immediate vicinity of the ignition source. Smoke and gas from this small fire would have been a serious threat to the occupants if allowed to spread through the building.

6.3.3.2

A group of teenage inmates of a state "training school" barricaded themselves in a recreation room in protest against disciplinary policies. The outside exit to the room was locked and windows were barred. The room was of non-combustible construction. The inmates piled furniture, including chairs and sofas upholstered with polymeric foam material, against the door leading to the rest of the building. One of the inmates set fire to the furniture to attract attention. Then they tried to suppress the fire by smothering and beating it out with materials at hand, but were unsuccessful. The fire grew rapidly in the pile of furniture, filling the room with smoke and gas. The fire alarm sounded and help arrived promptly, but by the time the door was forced open, the room was completely filled with smoke. The fire department quickly extinguished the fire, which was confined to the furniture. One inmate was dead from inhalation of toxic gases and the others required hospitalization. There was no significant structural damage to the building.

6.4 Vehicle Fires

6.4.1

A father picked up his son at college at the end of the spring term. The son's belongings were loaded in the back of the station wagon. As they were driving down the interstate highway, an odor of smoke was noticed. At first it was faint and was assumed to be from the heavy traffic. It grew stronger and a faint haze became visible in the rear view mirror. At this point, they pulled off onto the shoulder of the road and opened the back of the station wagon to investigate. Smoke was seen rising from a foam rubber pillow that had been placed on top of the other items. It was removed from the car and torn open, whereupon it burst into flames and was totally consumed at the roadside. The interior and contents of the car were saturated with the odor of burning rubber and required extensive cleaning. There was no other damage. The sun, shining in the curved rear view window of the station wagon during a long stretch of straight road, had h eated the foam rubber to the point where auto-oxidation started and heat built up within the well-insulated interior. When the pillow was removed from the car, the smoldering foam rubber was exposed to a fresh supply of air and open flaming resulted.

6.4.2

Children were left in a locked automobile while the parents went shopping. The children, playing in the back seat, found a pack of matches. The upholstery materials in the back seat became ignited, and then the foamed plastic padding material became involved. The children died of burns and smoke inhalation before the fire department arrived. Rescue was attempted by several passersby, but the heat was too intense.

6.4.3

A bus driver had just parked his new bus at a turn-around spot after unloading his passengers approximately two blocks before. He smelled smoke, and upon turning in his seat saw smoke coming from the rear of the vehicle. He attempted to extinguish the fire with a portable extinguisher, but was driven back by the intense smoke. The foamed plastic seating was the initial material observed to be on fire. Shortly thereafter, the vinyl wall coverings and synthetic plastic windows became involved. Flashover occurred in less than three minutes after discovery of the fire and the entire bus interior became involved. The interior of the bus was destroyed in less than five minutes, before the fire department could extinguish the fire. Fire officials investigated the fire and concluded that arson was its probable cause.

6.4.4

As a subway train approached an underground station, a blown rubber radial tire caused intermittent contact with the live third rail. This intermittent arcing occurred until the tires and hydraulic lines in the undercarriage caught fire. After these components ignited, the fire penetrated the plywood undercarriage and proceeded to burn the cabin interior, which consisted of plastic interior finish and foamed plastic seating. Passengers had been removed prior to the burn-through and a trainman attempted to extinguish the fire by directing a portable carbon dioxide extinguisher at the flames. Fire fighters were unable to approach the fire for over two hours due to the smoke and intense heat. Four cars out of ten were completely destroyed.

6.4.5

A disc brake on the underside of a modern subway car overheated and caused heating of an aluminum plate, which served as a barrier to external penetration. The car was unoccupied at the time. The aluminum plate covered a foamed plastic insulation material which became heated and then ignited. Open flaming occurred and was intensified by the rupture of hydraulic lines which served the brake. The fire was witnessed by a passerby when the train was above ground. The fire department was called and met the train at the next convenient area. In the intervening period, the fire had broken through the undercarriage and ignited the vehicle's seat cushions (foam plastic) as well as fiberglass reinforced plastic lining materials. Firemen wearing breathing apparatus entered the car and quickly extinguished the fire with hand lines. According to fire officials, the car appeared to have been on the verge of flashover when the fire department arrived. Extensive damage to the car occurred.

6.4.6

A ship was alongside a tender for upkeep and minor repairs. Damaged and worn out plastic foam mattresses were being replaced. Several of these items were briefly stacked against a metal bulkhead until the hatch could be cleared for their removal. An engineering technician was using a torch to cut metal support clips from the opposite of the bulkhead. He had properly requested, received and instructed a fire watch on his side of the bulkhead. No fire watch was ordered for the other side because there were no openings in the bulkhead was presumably devoid of combustibles. Yet, heat from the torch, transmitted through the metal bulkhead, ignited a mattress resting against it. The fire was discovered when smoke was seen pouring from an open hatch. Heavy black smoke hindered firefighting efforts. Fortunately, there were no combustibles other than the mattresses present to spread the fire. The fire was extinguished without serious structural damage, but with extensive smoke damage to the compartment of origin and adjoining compartments. Several crew members suffered from smoke and gas inhalation, but recovered without permanent injury.

6.4.7

Shortly after leaving a ship's lounge, a visitor on the vessel returned to the lounge to pick up an item he had forgotten. When he arrived, smoke was pouring through an open door. He immediately went to an alarm station and pulled the ship's general alarm. The lounge was tastefully provided with upholstered furniture, synthetic carpeting material, and vinyl wall covering. It is surmised that a cigarette caused ignition of a sofa which in turn, ignited the carpet that subsequently spread the fire to other furnishings. Rapid heat buildup resulted in flashover prior to arrival of the fire fighting team. The bulkheads and decks, which were of noncombustible construction, held the fire to a single compartment. The vessel's crew extinguished the fire. However, the lounge was a total loss. Fire fighting efforts were hampered by extremely acrid smoke. Subsequent investigation indicated that, in order to meet fire retardancy requirements, the foamed plastic utilized in the furnishings had been manufactured with a halogenated fire retardant. When subject to a high temperature environment, the fire retardant broke down into halogen acids.

6.4.8

While a ship was undergoing repairs, a torch was being used to cut metal hangers from the overhead of a living compartment. Globules of molten metal and slag fell to the deck and down a nearby open hatch, landing in a trash container in the compartment below. Although a fire watch was present, this threat to the lower compartment was not noticed. The trash container contents included paper, discarded electrical and communications wiere, wood scraps and plastic packing material from components just installed. Ignition was followed by rapid growth of the fire in the trash container. Flames engulfed the nearby bunk mattress and spread to the wiring bundle in the overhead. The fire spread to nearby compartments through open doors and hatches before it could be extinguished. Large amounts of black smoke were generated and spread through the area, entering the exhaust side of the ventilation systems. Smoke and fire damage were extensive in two compartments; with substantial fire spread along the wiring bundles into other areas. Personnel were affected by smoke and gas, but there were no fatalities.

6.5 Fires in Mines and Bunkers

6.5.1

A mine's electrical system had recently been renovated and new high voltage power lines were fitted underground. The previous terminal box specifically designed to be explosion proof, was utilized for the new wattage. Unfortunately, the PVC covered cable that was joined in this large 3′ X 4′ X 4′ junction box, required more ventilation to carry away the heat generated by the insulation material to break down to form an explosive gas mixture in the box. Subsequent high voltage arcing caused the heavy metal terminal box to explode with tremendous force, throwing chunks of metal 400′ along the shaft. One large section of the box severed emergency power cabling putting sources of light and ventilation in the shaft out of action. The explosion started a fire, which spread to a rubber covered conveyor belt. To extinguish the fire required extensive efforts over a two day period. The entire 400′ length of belt required replacement.

6.5.2

A urethane foam, formulated with a fire retardant, was being installed in a ventilation system of a coal mine shaft. The material was being foamed in place and its overall thickness was a nominal 6″. The equipment operator inadvertently utilized a mixture too rich in catalyst. After completing the block, he left the area and removed his equipment. Four hours later the excessive catalyst caused the urethane to self-heat and burst into flame. The fire was detected by smoke rising out of the ventilator shafts and a quick response was made. Teams of experienced fire fighters made their way to the entrance of the shaft and utilized the prevailing ventilation to speed the delivery of high expansion foam, which extinguished the fire.

6.6 Fires in Tents, Recreational Vehicles and Mobile Homes

6.6.1

While a housewife was preparing a meal in the kitchen of a mobile home, grease in a frying pan on the stove caught fire. She called for help, but, by the time her husband could respond, the foam plastic cabinet above the stove was involved. He brought a garden hose from outside, but by this time the flames were spreading over the plywood interior finish and he was unable to enter because of the heat and smoke. The fire department, called by a neighbor, responded within five minutes, but, by this time the trailer was completely involved, resulting in total loss.

6.6.2

Two nine-year-olds were sleeping in a pup tent in their back yard. They had built a fire near the open end of the tent earlier and had gone to bed when it burned down to glowing coals. During the night, a gust of wind blew a spark against the tent, starting a smoldering fire, which was fanned by wind and burst into flame. The fire spread rapidly over the water-repellent-treated fabric. The parents, sleeping were awakened by the children's cries. One child escaped through the burning open end of the tent and received severe burns when his clothing caught fire. The second child attempted to get out through the closed end of the tent away from the fire, but was trapped under the collapsing tent and died of burns. There was no fire spread to other structures.

6.6.3

Smoke grenades, set off near a field hospital unit to simulate an attack, set fire to dry gas. The fire spread rapidly due to 20 mph winds. An attempt was made to halt the fire at a road using shovels and other improvised equipment. Firefighting was hindered by a lack of water. The unit's water truck had been secured by chaining and locking the steedring wheel, but the driver with the key could not be located immediately. Within five minutes, the fire had broken across the road and reached a large supply tent. The tent burned rapidly scattering firebrands over the area and increasing the spread of the fire. A large inflatable shelter used as the unit's headquarters was the next structure to catch fire. It was constructed of plastic coated fabric, which was reported to have been flame retarded when manufactured. The unit was several years old. This inflatable shelter was completely destroyed in three minutes. Meanwhile, the fire spread to other structures. It was observed that the air locks at the entrance to the shelters would have been more severe hindrance to the evacuation of patients in an operational situation. Fortunately, there were no patients involved in this training exercise and no casualties resulted. A total of seven tents and inflatables, including their contents, were destroyed before base fire fighting units arrived and extinguished the fire.

6.7 Conclusions

All of the foregoing illustrative fires can be characterized by the generalized scenario as follows:

a. Combustible materials were deployed in a manner conducive to fire development.

b. An energy source was applied to an easily ignitable fuel element.

c. the fire grew and spread, consuming fuel and producing heat, smoke, and toxic products.

d. The fire was detected.

e. Fire control action was undertaken.

f. The fire was ultimately extinguished.

g. Loss resulted from the fire.

It should be evident that the fire could have been prevented, or the loss minimized, by more effective action at each step. Details of these fire scenarios help identify the areas where the most effective measures can be taken to minimize fire impact on similar situations in the future.

CHAPTER 7

ANALYSIS OF SPECIFIC SCENARIOS

7.1 Introduction

To illustrate the fire scenario approach to increase fire safety, two generalized fire scenarios are developed; one deals with an apartment fire, the other with an inflight aircraft fire. These scenarios, which include the essential fire elements (e.g., ignition source, first material ignited, etc.), permit generalization from the particular incident described. Moreover, these scenarios are analyzed to pin-point hazards, suggest opportunities for fire prevention, and direct attention towards methods for control. The analyses question the adequacy of specific critical materials, design approaches and regulatory codes and procedures; although the committee has excluded "review for adequacy of specific codes" from its responsibilities.

7.2 Apartment Fire

7.2.1 Scenario

A fire started in a lounge chair in the northwest corner of Mr. and Mrs. John Doe's living room. The Does live on the second floor of a recently constructed three floor condominium apartment building. Each floor contained four similar two bedroom apartments clustered about a stairwell. Construction, typically of wood, met local and FHA codes including those for fire walls, doors, etc. One course of brick veneer was applied to the outer walls. Access to the building was provided through a front hallway to an open stairwell. The second floor hallway had steps leading up and down. In addition to the steps leading down, the third floor hallway contained a ladder leading to an attic-type partial floor above it. Since only three floors above ground level were occupied, there were no fire escapes or other provision for secondary egress.

The furnishings, made entirely or partially or polymeric materials, included furniture, rugs, draperies, wall coverings, etc. The plumbing system was constructed of plastic pipe; the combination tub-shower stall was also made of plastic.

Mr. Doe had been watching the Monday night football game after enjoying a late dinner. Two cocktails, a pleasant dinner and a liqueur made the first half of the game more enjoyable. At halftime, Mr. Doe had another drink followed by a cigarette and then a second cigarette. Mr. Doe drowsed off while smoking the second cigarette; he was awakened by the noise of the crowd cheering a touchdown and decided to turn off his television set and go to bed.

The second cigarette had fallen from his hand into the chair. It smoldered there for 45 minutes before a flame appeared. The flame grew, going up the back of the chair and through the cushion. Heated pyrolysis and combustion gases rose to the 8 foot ceiling. /Burning intensity increased rapidly, spreading smoke and hot gases into other rooms.

Flashover occurred in 8 minutes (from flaming ignition) propagating flames into the bedrooms, spreading smoke and hot gases into the bedrooms, hallway, bath and kitchen. Hot combustion products and smoke poured through the front door, opened by Mrs. Doe during escape, into the center hallway and up the stairwell. Plastic fixtures in the bathroom quickly pyrolyzed, then burned. Smoke and flames followed the plastic plumbing system into the floor above, breaking out into the bathroom. The fire continued to grow on the second and third floors. It burned more intensely on the second floor, engulfing structural elements including the main stairwell. Firemen were called by an occupant of another appartment at this point, arriving simultaneously with the flashover of the upper stairwell. They rescued some of the occupants by ladders through windows, contained the blaze in 15 minutes, and put out all visible flames in less than two hours. Mop-up operations continued for several hours.

The Doe's apartment, and the one directly above it, were gutted. Downwind apartments on the second and third floors were heavily damaged (the wind was about 10 mph). Lesser damage was suffered by immediately adjacent apartments, but the ground floor, while undamaged by flames, did experience water damage.

Two elderly persons living on the third floor, over the Doe's apartment, died of smoke inhalation and/or burns. In the downwind third floor apartment, all four occupants, although affected by smoke, were rescued by firemen, as were the occupants of the adjacent apartment. Ground floor occupants escaped unharmed.

Since this scenario is plausible, its analysis in detail should reveal corrective measures applicable to similar occupancies and perhaps to general fire situations.

7.2.2 Analysis of the Apartment Fire

7.2.2.1 Pre-Fire

Analysis of a fire scenario should start with events that occurred long before the initial release of heat that sets the fire chain in motion. In some cases, an optimum solution to the fire problem evolves from action taken long before the fire starts. The analysis should then logically proceed through the various stages of the fire and end with a critique of the situation.

As a preliminary step, analysis requires the gathering of data relating to the origin and cause of the fire as well as other pertinent data such as legal documents, building codes, regulations, plans and specifications, builders' and manufacturers' records, inspection records and notes of oral discussions with interested parties.

Design Considerations

Legal and Regulatory

Analysis begins with a determination whether appropriate codes were specified, applied, and incorporated into the building plans and specifications. Improper application of codes or failure to comply with them could lead to civil and criminal litigation.

Review of the architect's and building contractor's records may disclose authorized or unauthorized alterations, substitutions or omissions that have the effect of nullifying code requirements.

Inspection records maintained by regulatory agencies may reveal patterns of marginal performance which, in total, provide unsatisfactory results. Lack of proper fire walls between apartments permits rapid lateral spread; inadequate apartment entrance doors permit fire spread and gas penetration from stairwells, inadequate fire stops in framing accelerate vertical fire progression through structures. Exposed charred remains, and neighboring undamaged structures, afford excellent bases for comparison of construction details with regulatory requirements.

The analysis of this apartment fire scenario found that the architect, builder and inspectors were knowledgeable and diligent in compliance with applicable regulations.

System Saety Considerations

System fire safety considerations are as important as code or regulatory items, which generally only relate to *component* requirements.

No system safety regulations or code items were applicable to the three story condominium apartment; no system safety review was made.

Impact of various fire sources and their probable consequence were not studied by the architect, builder or sales agency.

No escape plan was drawn; no analysis of the finished building for fire escape routes and related problems was made.

In-place survival potential for those cut off from escape routes was not studied.

The stairwell walls were covered with decorative panel board (3-ply wood board with a plastic finish); fire spread up the paneling could take place easily; third floor occupants would be trapped in their apartments even if they had the presence of mind to keep their doors closed; if not, the hot noxious gases would flow into those apartments with disastrous effect.

No heat or smoke detectors were installed in the apartments or stairwell. Smoke detectors in the stairwell or apartments would have provided early waning to apartment occupants not initially aware of the danger they were facing.

There were no sprinklers in any of the apartments or in the stairwell. Properly sensored sprinklers in the stairwell could have controlled fire spread in the exit path, providing substantial additional escape time.

Fire control methods were simplistic. Two 15 pound CO_2 extinguishers were available. One was located in the lower hallway and another in the furnace room. Fire hydrants were located just outside each building and at locations on the periphery of the parking lot.

The other condominium buildings in the cluster were well spaced in accordance with code and insurance requirements so that fire spread by jumping between buildings was unlikely. However, the parking lot arrangement was not satisfactory. The fire engines and ladder trucks were unable to get past the few row of cars. All hoses and apparatus had to be carried by the firemen until the small utility fire truck arrived. Modification of the parking area with no loss of parking spaces would have permitted direct fire engine access to the fire plugs peripherally located on the parking lot.

Summary

Application of a fire safety based analysis of the condominium and its modes of operation at the design, building and use stages, would have produced substantial fire safety improvements. Such analyses should be required for all new multiple dwellings.

7.2.2.2 Ignition Source

Scenario

The ignition source was a cigarette that fell into the over-stuffed chair.

Analysis

Not much can be done to reduce the ignition capabilities of today's carelessly discarded cigarettes. A type of cigarette exists that goes out after a short time if not smoked; it should be further developed since cigarettes are a frequent cause of fires. Even today's cigarette requires favorable conditions before it will ignite other objects. Many carelessly discarded cigarettes go out without igniting other materials. A continuing education program emphasizing the dangers of smoking while half asleep, in bed, or in a semi-intoxicated state, offers some promise, but major efforts should be directed toward countering the ignition source since this approach offers great potential for reducing fire incidents.

A cigarette is a serious ignition hazard. Because of its self-perpetuating glowing (not flaming) combustion, it can remain undetected in a smoker's environment for relatively long periods of time. This allows the fire a long period to develop and become deep seated. During such long induction periods, large quantities of lethal gases may be produced. Fires from other ignition sources, such as matches and lighters, are usually quickly detected and countermeasures can be started promptly.

Other ignition sources that could have initiated this type of fire (i.e., slowly developing) include "instant-on" televisions, electrical heaters, fireplaces, and overloaded or damaged wiring.

7.2.2.3 Ignited Material

Scenario

The cigarette was accidentally dropped into the space between the seat cushion and the arm of a typically upholstered overstuffed lounge chair. The cotton covering material smoldered at first, allowing the heat from the glowing cigarette (800 °C) to contact the urethane foam cushion.

Analysis

Within the present state-of-the-art of materials, there is much that can be done to improve the fire resistance of chair coverings and cushions. In fact, current development programs have already proposed cushion materials and covering materials of substantially higher resistance to ignition and flame spread. The National Bureau of

Standards is developing new testing methods that measure the response of materials to cigarette type ignition sources. Moreover, the Consumer Product Safety Commission is considering the promulgation of a mandatory standard which would eliminate easily-ignited upholstered furniture from the market place. Completion of this test program as well as the application of new materials and designs as appropriate would promote improved fire safety.

Fire safety for apartment occupants could be markedly increased by the installation of an inexpensive smoke detector, which would provide a warning during an early stage of the fire, probably during the smoldering period.

If the ignited material had been discovered earlier, the fire could have been contained rather easily by cutting out the affected portion and removing it, or the entire chair, from the building. A smoke detector and a properly sized fire extinguisher should be a part of the furnishings of every home.

The analysis might also consider other likely ignition sources and their locations in the apartment. Kitchens produce many fires from grease and appliances. Smoke detectors in kitchens are not useful, because of false alarms from cooking smoke. However, since kitchens are usually occupied when flames erupt, it is much better to have a fire extinguisher on hand. Kitchen trash containers often receive cigarette ashes and butts. Trash containers should be metal rather than a combustible material. In the past, bathrooms have been relatively fire safe, but the new plastic tubs and showers can contribute significantly to fire growth. Bedrooms have several fire sources including people who smoke in bed, children who may play with matches, appliances, wiring, etc. A hall smoke detector would provide some warning of fires from all these locations. Individual families, when assessing their living habits and accommodations, could develop priority requirements for the acquisition of fire detection and control equipment.

7.2.2.4 Smoldering and Flaming Combustion

Scenario

Although the cigarette, upholstery material and urethane cushion smoldered for about 45 minutes, they produced only small amounts of smoke and gas. Then, burning characteristics changed to full flaming combustion, rapidly engulfing the chair arm and cushion. Large quantities of hot smoke and gas developed and rose to the ceiling.

Analysis

Even with the given ignition source characteristics and system geometry, current covering and cushioning materials as well as designs, when properly employed, can significantly increase fire safety. Additionally, some exotic "space age" materials that are extremely fire resistant could be brought into common use through continued efforts to reduce their costs.

Control of a fire by apartment dwellers following the transition to flaming combustion is difficult. Since the fire had not been noticed until the flame broke out, the safest procedure would have been to leave the apartment, notify the other occupants of the building and call the fire department. This course of action was not available to the sleeping Doe family.

7.2.2.5 Flame Spread

Scenario

As flames broke out in the back of the chair, the polymeric window draperies caught fire. The fire spread to the wall paneling and nearby coffee table made of structural foam. Soon, the entire corner of the living room was blazing; producing copious quantities of hot smoke and noxious gases that rapidly filled the small plenum overhead and began to flow through the top of the living room entranceway into the hall and other rooms with open doors. Radiation from the hot gases and ceiling heated the other furnishings into the room to their ignition points and, shortly thereafter, flashover engulfed the living room.

Analysis

Review of this segment of the fire chain revealed several excellent possibilities for increasing fire resistance:

1. Replacement of the draperies with drapes made of more fire resistant material, e.g. fiberglass.
2. Replacement of the paneling with gypsum wall board.
3. Replacement of the coffee table and other plastic furniture with wooden furniture, or even more fire resistant plastic furniture.

Control of the fire at this stage demanded professional fire fighters and their equipment, but at this time in our scenario they had not been summoned.

7.2.2.6 Evolution of Smoke and Toxic Gases

Scenario

See 7.2.2 through 7.2.2.5.

Analysis

Smoke and toxic gases evolved in each of the phases described earlier. Smoldering combustion generally produces more CO and less heat than does flaming combustion; smoke quantities vary considerably.

smoke detectors are available in a wide range of capabilities and price. Fortunately, relatively inexpensive, but quite effective, detectors are available. Heat detectors are also available at a wide range of price, but they are better suited to property protection than to life-safety applications, since they are *less rapid in their response*. Reliable and inexpensive detection of the toxic gases, principally CO, is not yet available.

Visibility is markedly affected by smoke. Smoke obscures or obliterates familiar landmarks and causes some distortion; it irritates the nose, mouth and throat. Toxic gases and smoke affect the eyes (watering) and reduce visual efficiency in many other ways. Toxic gases have severe effects on the brain, nervous system, respiratory system and heart. Although it has not yet been quantitatively determined, there is some qualitative evidence that combinations of toxic gases interact with humans in a way such as to multiply their individual effects. The toxic gases generated in this scenario would have been sufficient to produce the death of all members of the Doe family.

In a fire, thermal burns are occasionally the cause of death. More frequently, as in our fire scenario, thermal burns on the outer skin or in the respiratory system are not the primary cause of fatalities. Smoke and toxic gases incapacitated the victims, preventing their escape. The burns suffered by Mr. Doe were a secondary effect.

The analyses presented in 7.2.2.2 through 7.2.2.5 have suggested ways to reduce the probability of fire, or to diminish its intensity, by reducing the volume and effects of smoke and toxic gases. Humans can take rudimentary but effective steps to reduce the adverse effects of smoke and toxic gases; these means include:

1. remaining close to the floor while waiting or moving.
2. moving from the contaminated area as soon as a safe course of action is planned.
3. remaining in a quiescent posture, with minimum exertion and breathing, except when necessary to move for safety or escape.
4. covering the nose and mouth with wetted cloths (ineffective against carbon monoxide).
5. maintaining a barrier between the fire source and a temporary haven (i.e., keep the room door closed unless opening is mandatory); close cracks around the door with a damp cloth.

While implementation of these steps is largely a matter of education and training, very few of these steps are practiced or even recognized as contributing to fire safety, with the exception of fire drills performed by school children. To make matters worse, standards for the control of smoke and toxic gases in residential buildings are not under development except as by-products of fire control methods.

7.2.2.7 Flashover

Scenario

Shortly after flaming combustion engulfed part of the living room, the superheated pyrolysis products filled the upper levels of the hallway, kitchen, dining area, and bedroom through the open doors. Flashover occurred initially in the living room. Mrs. Doe awakened at this point, grabbed her child and was just able to leave the apartment. Smoke and gases rapidly filled the bedrooms, flames moved laterally along the upper walls supported by new air coming in at the bottom of the main entrance doorway. Hot gases filled the upper part of the bathroom and pyrolyzed the plastic shower walls, which then fed still more hot gases. Flashover followed shortly thereafter and the plastic tub burst into flame, sending hot gases up the plumbing shaft into the bedroom of the apartment above. One by one the bedrooms flashed over and were engulfed by flames.

Analysis

Once conditions for flashover develop, full scale fire fighting efforts are required to stop the process and protect adjacent structures.

7.2.2.8 Spread to Adjacent Compartments

Scenario

Hot smoke and gases flowed from the Doe's apartment into the hall and stairwell, rapidly filling the third level. Flames exited the Doe's door and ignited the walls of the hallway; flashover occurred on the third level in a matter of minutes, gradually engulfing the hallway in flames from the second to the third floor. Simultaneously, the fire broke through the bathrooms and the living room into the third floor directly above. The elderly occupants of this apartment were unable to escape. On the other side of the condominium building, occupants were unaware of the fire until the stairwell flashed over. The third story occupants were rescued by firemen, then the second story occupants were rescued. All first story occupants, aroused by flashover noises and the fire department's arrival, were rescued or exited unassisted. Meanwhile, the fire spread through structural walls, fed by polymeric foamed insulation.

Analysis

No fire escapes were required by code, therefore, none were provided. None of the third floor occupants had installed portable ladders for escape.

Fortunately the windows and storm sash could be opened for escape. Sealed windows and certain casement, jalousie or awning type windows make escape difficult or impossible. Special escape routes planned by the architect may be required under such circumstances.

Full sprinkler installation in all portions of the condominium unit would clearly add to fire safety. Such installations may be too costly for some types of housing. A sprinkler system for the common use stairwell, and particularly for the furnace room, would offer major escape improvement possibilities at a much lower cost.

At this phase of the fire chain, control and extinguishment requires fire professionals and their equipment; the analysis does not reveal substantial opportunity for improving fire safety except through optimum use of improved materials, self-closing doors, early detection of the fire, and prompt notification to fire fighters.

7.2.2.9 Spread to Other Structures

Scenario

The fire was contained in the condominium unit where it started.

Analysis

Although not within the province of this committee's charge, an analysis should proceed beyond simple acceptance of the lack of fire progression into the neighboring structures. Key factors should be examined, such as:

1. Are the structures in line so that a prevailing or constant wind could propagate fire unusual distances, or spread burning fragments in such directions? (This possible circumstance suggests a "worst case" analysis).

2. Can the water mains in the area support fighting more than one fire at a time?

3. Do firemen have proper access to all units for fire fighting and occupant rescue when parking lots are full and when conditions are unusual (such as heavy snow banks caused by plowing)?

7.2.2.10 Extinguishment

Scenario

Neither the Doe family nor any other occupants of the condominium made any effort to extinguish the fire. Firemen on arrival applied conventional fire fighting techniques to quickly control and extinguish the fire.

Analysis

Early extinguishment of the smoldering chair would have drastically changed this scenario. Once the apatment was ablaze, the fire department was required.

7.2.2.11 Summary

Analysis of this hypothetical scenario led to several recommendations that, if implemented, would be effective in reducing the adverse effects of this type of fire, at relatively low cost, for the specific types of dwelling units examined. These recommendations include:

1. Require upholstered furniture to have a covering material-foam cushion combination resistant to a cigarette ignition source.

2. Each apartment and common hallways should be equipped with smoke detectors for early warning.

3. Sprinklers should be installed in hallways and stairwells.

4. Building design and construction practices should include provisions for fire safety. Education of architects, engineers and builders in fire safety would be particularly beneficial.

5. Develop an acceptable cigarette which will self extinguish after a short time if not smoked.

(Note: only recommendation (1) above falls within the committee's immediate materials interest.).

7.3 Aircraft In-Flight Fire

7.3.1 Scenario

In this hypothetical scenario a transatlantic commercial jet airplane, nearing the end of its flight, was approximately 30 minutes from its destination. A passenger, upon entering a closed lavatory at the rear of the aircraft, encountered a wave of white smoke. He quickly closed the lavatory door and called for the attendants.

The first steward went into the cockpit and reported smoke and fire in the aft lavatory. A cabin crew member discharged two fire extinguishers into the lavatory, but smoke continued to increase and become darker.

The captain reported the fire to ground control and requested clearance for emergency descent. At the same time, he ordered the cabin depressurized and sent the flight engineer back to analyze the situation. The flight engineer went to the rear of the aircraft taking with him a CO_2 fire extinguisher bottle. When the flight engineer saw black smoke filling the area behind the last row of seats, he handed the extinguisher to a steward and quickly was descending, the flight engineer increased the airflow to the cabin to keep the smoke in the rear of the airplane.

Soon thereafter, a steward came into the cockpit reporting that the cabin was half filled with smoke and passengers were being affected. The captain ordered an over-wing emergency window removed. A steward, equipped with an O_2 bottle and a full face mask, tried unsuccessfully to comply with that order.

Approximately 3 minutes after the first report of smoke and fire in the lavatory, smoke reached the cockpit. Visibility was reduced so that the pilots could neither see their instruments nor see through the windshield. Both pilots opened their sliding windows. Visibility was possible through the open windows as the flight continued.

The captain decided to land as soon as possible and landed the aircraft soon thereafter in an open field. During the landing both main landing gears broke off. The fuselage came to rest practically intact. After the aircraft came to rest, the fire, already in progress with the fuselage, broke out through the top of the cabin in front of the vertical fin. This fire consumed virtually the entire fuselage interior. Only one passenger and some crew members survived.

7.3.2 Analysis of the Aircraft In-Flight Fire

7.3.2.1 Pre-Fire

Design Considerations

Legal and Regulatory

The aircraft was constructed by an established manufacturer, competent in every respect, who complied fully with applicable regulations. Operating procedures were also in accordance with regulations. The crew was well trained and fully experienced. The aircraft had been properly serviced and loaded.

System Safety Considerations

Government Regulations

Federal Aviation Administration regulations prescribe in considerable detail the types of materials that may be employed for specified end uses in aircraft, the tests that must be applied, and, in general, all matters relating to public safety including modes of operations.

Analysis indicated that the state of the art did not provide guidance regarding systems applications of materials and their systems interrelated behavior, nor the operational aspects of systems interrelationships. There was no methodology for determination of risk, assessment of risk consequences or risk-cost trade off. Under such circumstances, despite the intelligent use of available data, materials, and methodology, major deficiencies in fire safety performance existed.

Material Considerations in Aircraft Design

As in other complex systems, aircraft are designed in accordance with regulations to meet mandatory functional requirements. When these regulations are met, the mandatory safety requirements (including fire safety) are assumed to be incorporated in the design. Yet, review and analysis of the aircraft design from a fire safety standpoint revealed:

1. That fire safety considerations could have included more comprehensive attention to system interactions.
2. That material and material test requirements could be better established on the basis of material interactions and fire dynamics in realistic complex environments.
3. That existing risk analysis techniques could be improved and augmented to enable designers to conduct more comprehensive trade off studies of individual components and their relationships to total system performance.
4. That improvement in fire safety in the cabin had been attempted without adequate consideration of the potential smoke and toxic effects of the materials employed.
5. That there was an insufficient fire safety technology base for the FAA and aircraft manufacturer to confidently direct or perform actions to rectify the major deficiencies implied above.

Aircraft Manufacture

The procedures used in manufacturing this aircraft, and others on the assembly line, were proven by many years of successful experience. This fact was further confirmed by quality assurance methodology. Initial certification revealed no deficiencies.

Aircraft Operations

The operating organization was long established and had an excellent safety record. Maintenance and operating procedures were well defined and written. The training of employees was thorough. Inspection of aircraft facilities, maintenance and operational procedures by government agencies disclosed a few minor deficiencies, which were quickly corrected.

Fire Safety Aspects

All required fire safety analyses, tests and demonstrations, including the evacuation demonstration, were complied with by the manufacturer and operator.

Summary

The aircraft fully complied with FAA regulations.

7.3.2.2 Ignition Source

Scenario

The fire started from a cigarette dropped into the paper towel disposal slot. It landed on a dry paper towel and a film wrapper from a pack of cigarettes.

Analysis

Despite a ban on smoking in lavatories, some persons ignore it. They drop ashes into the wash basin and cigarette butts (hopefully extinguished) into waste containers out of ignorance or to conceal the violation of the no-smoking rule. Effective enforcement, possibly using smoke detectors, would eliminate this problem.

7.3.2.3 Ignited Material

Scenario

The material ignited by the cigarette was the cigarette package wrapper and paper toweling discarded through the waste disposal slot into the waste bin. Either could have supported ignition, although waste paper toweling often smolders and extinguishes. Since paper is a proper product for the dispenser, disposal arrangements should be fire hardened. The waste bin was plastic. One step in fire safety would be to replace the plastic bin with a non-combustible bin.

7.3.2.4 Smoldering and Flaming Combustion

Scenario

The ignited cigarette package wrapper flamed. The discarded paper toweling smoldered and then flamed involving more paper. Soon the plastic trash bin began to pyrolyze and flame appeared, travelling upward, involving the plastic paneling of the lavatory framework.

Analysis

The waste bin could contain all kinds of combustible materials other than used toweling. Thus, either smoldering or flaming combustion could occur even from a low heat flux ignition source. Corrective measures require more than just fire hardening the trash container. Although system design per se is not within the committee's expertise, it appears that water application to the trash, or occasional inereting by nitrogen or carbon dioxide, offers worthwhile safety advantages. Another safety precaution would be the presence of a small self-actuating fire extinguisher.

7.3.2.5 Fire Spread

The fire spread up the plastic paneling, penetrating the lavatory overhead, and began to progress laterally, fore and aft. Smoke filled the adjancet lavatories, moving forward in the overhead plenum and into the passenger cabin. As the fire grew, additional smoke and gases poured into the cabin and moved forward into the cockpit, obscuring the vision of the crew.

Analysis

Once the waste bin and lavatory started flaming combustion, there was little the crew could do to get to the burning materials with the fire extinguishers (CO_2 portables). The openings above, below, and between lavatories facilitated the fire's movement. The make up-ventilation system that recirculated a part of the cabin air assisted the forward movement of the smoke and fire. Fire hardening of the lavatory compartment, using improved materials, would result in a higher degree of resistance to ignition and flame spread.

7.3.2.6 Evolution of Smoke and Toxic Gases

Scenario

The first smoke visually detected as coming from the lavatory was light in color, possibly from partially wet discarded paper toweling. As the fire progressed, the smoke rapidly became dark and dense. As it moved forward in the cabin, passengers in the cabin moved forward to vacant seats. However, they were gradually engulfed by the smoke, lost their vision, and were incapacitated. Depressurization of the cabin did not decrease the forward flow of smoke and noxious gases. Visibility was reduced to zero in the aft cabin and smoke entered the flight station, filling it, and seriously impairing the pilot's vision. The pilots, using oxygen masks for breathing, opened the sliding windows in the cockpit to restore visibility. Toxic effects, while not accurately known, were severe. After landing, only one passenger was able to escape through the opened galley door.

Analysis

Lack of smoke detectors in the lavatory delayed the discovery of the fire. Smoke and toxic gases were the primary factors in the heavy loss of life, although if the plane had remained in the air for a few more minutes, the fire would probably have caused a fatal crash.

The passenger oxygen system, required by regulation for depressurization situations, is a "make-up" system blending oxygen with cabin air to supply each passenger mask. Use of the passenger oxygen system would have distributed the smoke and toxic gases more rapidly to all parts of the cabin. Increased cabin oxygen would also have resulted in increased flame spread and enhanced transition from smoldering to flaming combustion. Oxygen masks provide no protection for the eyes. The FAA hood with individual oxygen supply (under development) would have provided each passenger with a survival system until the plane landed, and a substantial number of passengers might have survived.

There are no test data that reveal the direction of normal or casually induced air flow through the large overhead plenum that runs the full length of the aircraft. (The plenum has been increased in volume in wide bodied jets and the "wide body" conversions of older jet aircraft.) In our scenario aircraft, there was no compartmentation or design effort to utilize this space for controlling fire, smoke or hot gases. Actually, this space provided an uncontrolled contribution to the tragedy. Consequently, there is a clear need to develop a compartmentation and air flow control for this vital upper area. In this connection a fire in the forward galley or forward lavatory could have had similar end results by having the smoke and toxic gases distributed through the ventilation system.

The pilots and crew in the cockpit survived because they used portable oxygen masks and/or opened the windows. The smoke and toxic gases could only have entered the cockpit if it were at a lower pressure than the cabin. Provision should be made to assure that the cockpit is under slight positive pressure relative to the main cabin under any normal or emergency condition and that air flow is always out of the cockpit.

7.3.2.7 Flashover

Scenario

Although there were sufficient heat flux and hot pyrolyzed gases from the fire, flashover did not occur while the aircraft was airborne. After the emergency landing, the fire-gas development was increasingly intense. A series of flameovers then occurred, ultimately engulfing the entire passenger cabin (flashover). The fuselage was gutted, with upper structure and skin members melted away from the intense heat.

Analysis

Conditions for flashover were present; the configuration of the lavatories and galley were geometrically conducive to flashover. Venting of hot gases into the cabin volume diluted the heat and gas content as did the venting through aft outflow, thus delaying flashover in flight. Considering the explosive forces and almost instantaneous fire spread typical of flashover, it is probably fortunate that the smoke and hot gases were diluted by cabin air. Cabin flashover and the resulting severe cabin fire would have killed all the passengers and crew and caused the plane to crash earlier. There are no available data relative to the desirable tactics of fighting a large scale in an aircraft while in flight. There are no data to support venting to prevent flashover or compartmentation to contain flashover. Experiments and large scale tests are needed, utilizing various materials with improved fire safety properties.

Once on the ground, rapid buildup of heated gases from the burning interior filled the cabin and the first flameover occured. With continuing flow of air through the open door forward galley door and partially opened forward main passenger door, several smaller additional flameovers developed. At this point, with the passengers incapacitated and the aircraft systems inoperable, it made little difference that the fire spread more rapidly than in a full vented situation.

7.3.2.8 Spread to Other Structures

Scenario

The plane landed in an open field. No other structures were nearby.

Analysis

It is fortunate that the captain was able to land the aircraft under controlled conditions in a field. Another possibility that might have occurred during the earlier portion of this in-flight fire scenario, is a fire-induced explosion of the plane in midair with flaming segments falling to ground and crashing into buildings.

7.3.2.9 Extinguishment

Scenario

Extinguishment of this lavatory fire, as attempted by using CO_2 portable extinguishers, was unsuccessful. After the plane crashlanded in the field, no extinguishment facilities were available. The fire burned itself out.

Analysis

The unsuccessful attempt to extinguish the fire in the lavatory indicated that the fire was either sufficiently concealed or had progressed beyond the point where the CO_2 portable extinguishers could be effective. No other facilities were available to put out the fire. Extinguishment on the ground under the circumstances was not possible.

7.3.2.10 Summary

The fire safety aspects of this in-flight aircraft fire identifies a number of specific needs and recommendations as follows:

1. the need for a better fire safety technology base, particularly as it relates to materials performance in standard tests and their behavior in actual fire environments.
2. A need for smoke and/or thermal detection devices in lavatories.
3. The need for fire extinguishment capabilities in lavatories.
4. A need for the fire hardening of lavatories and galleys by selection of more fire resistant materials.
5. The need for smoke and toxicity standards for materials employed in aircraft interiors.
6. A need for controlled ventilation of the cabin during an in-flight fire.
7. A need for optimized procedures for coping with in-flight fires.

Compilation and review of the results of analyses of many aircraft scenarios would provide a better basis for developing improved fire safety. Specifically, such action would lead to:

- Selection of improved materials for use in aircraft.
- Development of improved small scale tests for aircraft subsystems and components, as well as materials to be used therein.
- Development of necessary full scale fire tests of aircraft.
- Development of improved aircraft fire safety design criteria and methods.
- Development of risk assessment and trade-off methodologies needed, but not now available.
- Development of improved aircraft operating and maintenance procedures.
- Development of optimized procedures for coping with in-flight fires.

7.4 Scenario Analysis

Scenario analysis is a powerful, underutilized tool in our efforts to improve fire safety. Like the "Case Method" in business studies, it permits both the broad or narrow focus of many experts on a defined (perhaps hypothetical) situation. It permits a logical sequence for proceeding from specific to general theorems of behavior in our environment.

Scenario analysis consistently confirms, as it did in the two preceding examples, that our fire safety development efforts are inadequately supported and, unfortunately, are largely directed to items or components, when, in fact, our problems are those of complex systems.

Polymeric materials are major elements in these complex systems; new materials are being developed and introduced without full understanding of the potential consequences of their use, particularly as they relate to fire safety performance interrelationships. Scenario analysis can assist in better defining the roles that materials can play and what the consequences of their uses may be.